Richard Lydekker

A hand-book to the Marsupialia and Monotremata

Richard Lydekker

A hand-book to the Marsupialia and Monotremata

ISBN/EAN: 9783741118999

Manufactured in Europe, USA, Canada, Australia, Japa

Cover: Foto ©Thomas Meinert / pixelio.de

Manufactured and distributed by brebook publishing software (www.brebook.com)

Richard Lydekker

A hand-book to the Marsupialia and Monotremata

LLOYD'S NATURAL HISTORY.

EDITED BY R. BOWDLER SHARPE, LL.D., F.L.S., &c.

A HAND-BOOK

TO THE

MARSUPIALIA

AND

MONOTREMATA.

BY

RICHARD LYDEKKER, B.A., F.R.S.,

Vice-President of the Geological Society, etc., etc., etc.

LONDON
EDWARD LLOYD, LIMITED,
12, SALISBURY SQUARE, FLEET STREET.
1896.

PREFACE.

THE volume which Mr. Lydekker has written will, I believe, be found to be very useful to the student of Mammalia. The volumes of Gould's "Mammals of Australia" are both rare and costly, and the object of the "Naturalist's Library" is to give, within a handy compass, a scientific, and yet popular, account of the Australian Mammals. For the scientist, Mr. Oldfield Thomas's "Catalogue of the Marsupialia and Monotremata in the collection of the British Museum" is the indispensable compendium of our knowledge of these two Orders up to the year 1888. During the six years which have since elapsed, a few new species have been described, and these have been added in the present work by Mr. Lydekker, whose notes on the fossil species, on which he is one of the first authorities, lend an additional interest to the volume.

Especial care has been bestowed upon the plates. Those originally published in "Jardine's Naturalist's Library" were very well executed, and in the present issue the whole of the animals figured have been re-coloured from actual specimens in the British Museum, while figures of some of the remarkable forms discovered since the first publication of the "Naturalist's Library" have been added. The woodcuts have also been re-drawn from actual specimens. The Editor acknowledges with pleasure the pains which Mr. Lydekker has taken to give a full account of the two Orders contained in the present volume, and also the liberality of the publishers in the matter of the illustrations.

R. BOWDLER SHARPE.

INTRODUCTION.

In few, if in any group of Mammals, is our knowledge of species and genera in such a satisfactory condition as is the case with the Marsupials and Monotremes. This is entirely due to the labours of Mr. Oldfield Thomas, of the British Museum, whose results are published in the "Catalogue" so often referred to in the course of the following pages.

Under such circumstances, the task of writing a popular monograph of the groups in question has been a comparatively easy one; and the present volume must be regarded as little more than an abridgment of Mr. Thomas's work put into a more popular guise, with the omission of certain osteological and other more technical details. Since the publication of his "Catalogue" in 1888, a few new species have been discovered or described, which are inserted in their proper places in this book, by far the most interesting of these being the Marsupial Mole, representing a type of the Order previously quite unknown. Moreover, since the past history of the Marsupials has an important bearing on their present very restricted geographical distribution, it has been felt that no treatise on the group would be complete without some reference to this subject. Accordingly, a section of the work has been devoted to the consideration of some of the more generally interesting extinct representatives of that Order, while it also includes the little that is known of the geological history of the Monotremes, and likewise takes cognisance of a remarkable extinct group believed to have some affinity with the latter. In a work of the present nature it would, however, obviously be inadvisable to treat of the fossil forms in the detail with which their recent allies are described, and, accordingly, nothing more than a general sketch has been attempted.

In marked contrast to the fulness of our knowledge of the Marsupials and Monotremes from the point of view of the systematist, is our deficiency with regard to their habits and mode of life, this being especially the case in respect to the breeding-habits of the Monotremata. It is not to the credit of the present generation that the working zoologist has for the most part to rely for his knowledge of the habits of the greater number of Marsupials upon observations—admirable enough in their way—published many years ago, and the Author cannot but hope that the appearance of the present little volume may act as a stimulus to those having the opportunities of increasing our knowledge on this subject, information being most specially desirable with regard to the mode of life of the smaller members of the group.

R LYDEKKER.

SYSTEMATIC INDEX.

	PAGE
ORDER MARSUPIALIA	1
FAMILY MACROPODIDÆ	11, 254
I. MACROPUS, Shaw	14, 254
1. giganteus (Zimm.)	15
2. antilopinus (Gould)	19
3. robustus, Gould	20
4. isabellinus (Gould)	22
5. rufus (Desm.)	22
6. magnus, Owen	24
7. ualabatus (Less. and Garn.)	24
8. ruficollis (Desm.)	26
9. greyi (Gray)	29
10. dorsalis (Gray)	29
11. parryi, Bennett	30
12. irma (Jourd.)	32
13. agilis (Gould)	32
14. coxeni (Gray)	34
15. stigmaticus (Gould)	35
16. wilcoxi (McCoy)	36
17. brunii (Schreber)	36
18. browni (Ramsay)	37
19. thetidis (Less.)	38
20. eugenii (Desm.)	39
21. parma, Waterh.	40
22. billardieri (Desm.)	40
23. brachyurus (Q. and G.)	41
II. PETROGALE, Gray	42
1. penicillata (Gray)	43
2. lateralis, Gould	44
3. brachyotis (Gould)	45
4. inornata, Gould	46

SYSTEMATIC INDEX.

	PAGE
PETROGALE, Gray—(continued).	
5. concinna, Gould	47
6. xanthopus, Gray	47
III. ONYCHOGALE, Gray...	48
1. unguifera (Gould)	49
2. frenata (Gould) ...	50
3. lunata (Gould) ..	51
IV. LAGORCHESTES, Gould	52
1. conspicillatus, Gould	52
2. leporoides (Gould)	53
3. hirsutus, Gould	54
V. DORCOPSIS, Schl. and Müll. ...	55
1. muelleri (Schl. and Müll.)	56
2. luctuosa (D'Albert.)	56
3. macleayi, Mikl. Macl. ...	57
VI. DENDROLAGUS, Schl. and Müll.	58
1. lumholtzi, Collett	58
2. ursinus, Schl. and Müll.	60
3. inustus, Schl. and Müll.	60
4. dorianus, Ramsay	61
VII. LAGOSTROPHUS, Thomas ...	61
1. fasciatus, (Péron and Lesueur)...	61
VIII. POTOROUS, Desm. ...	63
1. platyops (Gould)	64
2. gilberti (Gould) ...	64
3. tridactylus (Kerr)	65
IX. CALOPRYMNUS, Thomas	66
1. campestris (Gould)	66
X. BETTONGIA, Gray ...	67
1. lesueuri (Q. and G.)	67
2. cuniculus (Ogilby)	69
3. gaimardi (Desm.)	69
4. penicillata, Gray	70
XI. ÆPYPRYMNUS, Garrod	71
1. rufescens (Gray)	71
XII. HYPSIPRYMNODON, Ramsay	73
1. moschatus, Ramsay	73
FAMILY PHALANGERIDÆ	75
XIII. PHASCOLARCTUS, Blainv.	78
1 cinereus (Goldf.)	79

SYSTEMATIC INDEX.

	PAGE
XIV. PHALANGER, Storr	80
1. ursinus (Temm.)	81
2. maculatus (Geoffr.)	82
3. orientalis (Pall.)	84
4. ornatus (Gray)	86
5. celebensis (Gray)	86
XV. TRICHOSURUS, Less.	87
1. vulpecula (Kerr)	88
2. caninus (Ogilby)	91
XVI. PSEUDOCHIRUS, Ogilby	92
1. lemuroides (Collett)	93
2. herbertensis (Collett)	93
3. peregrinus (Bodd.)	94
4. occidentalis, Thomas	96
5. cooki (Desm.)	96
6. archeri (Collett)	97
7. albertisi (Peters)	98
8. schlegeli, Jentink	99
9. canescens (Waterh.)	99
10. forbesi, Thomas	100
XVII. PETAUROIDES, Thomas	100
1. volans (Kerr)	101
XVIII. DACTYLOPSILA, Gray	102
1. trivirgata, Gray	103
2. palpator, A. Milne-Edw.	103
XIX. PETAURUS, Shaw	104
1. sciureus (Shaw)	104
2. breviceps, Waterh	105
3. australis, Shaw	106
XX. GYMNOBELIDEUS, McCoy	110
1. leadbeateri, McCoy	111
XXI. DROMICIA, Gray	111
1. concinna, Gould	112
2. nana (Desm.)	113
3. caudata, A. Milne-Edw.	115
4. lepida, Thomas	115
XXII. DISTÆCHURUS, Peters	116
1. pennatus (Peters)	116
XXIII. ACROBATES, Desm.	117
1. pygmæus (Shaw)	117
2. pulchellus, Rothsch.	119

SYSTEMATIC INDEX.

	PAGE
XXIV. TARSIPES, Gerv. and Verr.	119
1. rostratus, G. and V.	120
FAMILY PHASCOLOMYIDÆ	122, 265
XXV. PHASCOLOMYS, Geoffr.	124, 265
1. mitchelli, Owen	125
2. ursinus (Shaw)	125
3. latifrons, Owen	128
FAMILY PERAMELIDÆ	128
XXVI. PERAGALE, Gray	132
1. lagotis (Reid)	132
2. leucura, Thomas	134
XXVII. PERAMELES, Geoffr.	135
1. bougainvillii, Q. and G.	137
2. gunni, Gray	139
3. nasuta, Geoffr.	140
4. longicauda, Peters and Doria	140
5. raffrayana, Milne-Edw.	141
6. broadbenti, Ramsay	142
7. doreyana, Q. and G.	142
8. cockerelli, Ramsay	143
9. macrura, Gould	143
10. aurata, Ramsay	144
11. obesula (Shaw)	144
12 moresbyensis, Ramsay	145
XXVIII. CHŒROPUS, Ogilby	146
1. castanotis, Gray	146
FAMILY DASYURIDÆ	150, 267
XXIX. THYLACINUS, Temm.	150, 267
1. cynocephalus (Harris)	151
XXX. SARCOPHILUS, F. Cuv.	153, 267
1. ursinus (Harris)	154
XXXI. DASYURUS, Geoffr.	157
1. maculatus (Kerr)	158
2. gracilis, Ramsay	160
3. viverrinus (Shaw)	161
4. geoffroyi, Gould	164
5. hallucatus, Gould	165
6. albopunctatus, Schl.	165

SYSTEMATIC INDEX. xi

	PAGE
XXXII. PHASCOLOGALE, Temm.	166
1. cristicauda (Krefft)	167
2. apicalis, Gray	167
3. thorbeckiana, Schl.	168
4. wallacii (Gray)	169
5. doriæ, Thomas	170
6. dorsalis, Peters and Doria	170
7. swainsoni, Waterh.	171
8. minima (Geoffr.)	171
9. flavipes, Waterh.	172
10. minutissima (Gould)	174
11. longicaudata, Schl.	174
12. penicillata (Shaw)	175
13. calura, Gould	176
XXXIII. SMINTHOPSIS, Thomas	177
1. crassicaudata (Gould)	178
2. murina (Waterh.)	179
3. leucopus (Gray)	180
4. virginiæ (De Tarrag.)	180
XXXIV. ANTECHINOMYS, Krefft	181
1. laniger (Gould)	182
XXXV. MYRMECOBIUS, Waterh.	182
1. fasciatus, Waterh.	184
FAMILY NOTORYCTIDÆ	187
XXXVI. NOTORYCTES, Stirling	187
1. typhlops, Stirling	189
FAMILY DIDELPHYIDÆ	193, 271
XXXVII. DIDELPHYS, Linn.	196, 271
1. marsupialis, Linn.	196
2. opossum, Linn.	200
3. nudicaudata, Geoffr.	201
4. crassicaudata, Desm.	202
5. philander, Linn.	205
6. lanigera (Desm.)	206
7. cinerea, Temm.	207
8. murina, Linn.	209
9. canescens, Allen	210
10. lepida, Thomas	211
11. pusilla, Desm.	211
12. grisea, Desm.	212

	PAGE
DIDELPHYS, Linn.—(continued).	
13. velutina, Wagner	213
14. elegans, Waterh.	214
15. dimidiata, Wagner	215
16. brevicaudata, Erxl.	216
17. domestica, Wagner	217
18. scalops, Thomas	217
19. henseli, Thomas	218
20. sorex (Hensel) ...	218
21. americana (Müll.)	219
22. iheringi, Thomas	220
23. unistriata, Wagner	220
XXXVIII. CHIRONECTES, Illiger	221
1. minimus (Zimm.)	221

ORDER MONOTREMATA ... 224

FAMILY ORNITHORHYNCHIDÆ .. 230
XXXIX. ORNITHORHYNCHUS, Blumenb. .. 231
    1. anatinus (Shaw) ... 231

FAMILY ECHIDNIDÆ 240
XL. ECHIDNA, Cuvier ... 241
    1. aculeata (Shaw) ... 241
XLI. PROECHIDNA, Gerv. ... 245
    1. bruijni (Peters and Doria) 246
    2. nigro-aculeata, Rothschild ... 247

FOSSIL MARSUPIALIA 249

FAMILY MACROPODIDÆ .. 251
A. PALORCHESTES, Owen ... 251
    1. azael, Owen ... 251
B. PROCOPTODON, Owen... ... 252
    1. goliah (Owen) 253
    2. rapha, Owen 253
    3. otuel (Owen) 253
C. STHENURUS, Owen ... 253
    1. atlas (Owen) ... 253
D. MACROPUS, Shaw 254
    1. titan, Owen 254
    2. ferragus (Owen).. ... 255
    3. altus (Owen) 255

SYSTEMATIC INDEX.

	PAGE
D. MACROPUS, Shaw.—(*continued*).	
4. cooperi (Owen) ...	... 255
5. brehus (Owen) ...	... 256
6. rœchus (Owen) ...	... 257
7. anak, Owen ...	... 257
8. minor (Owen) ...	... 258
E. TRICLIS, De Vis ...	... 258
1. oscillans, De Vis	... 258
FAMILY PHALANGERIDÆ ...	... 258
F. THYLACOLEO, Owen ...	... 259
1. carnifex, Owen ...	... 259
FAMILY DIPROTODONTIDÆ	... 261
G. DIPROTODON, Owen ...	... 261
1. australis, Owen ...	... 261
FAMILY NOTOTHERIIDÆ ...	... 263
H. NOTOTHERIUM, Owen	... 263
1. mitchelli, Owen...	... 263
FAMILY PHASCOLOMYIDÆ ...	... 265
I. PHASCOLONUS, Owen ...	... 265
1. gigas (Owen) ...	... 265
J. SCEPARNODON, Owen...	... 266
1. ramsayi, Owen ...	... 266
FAMILY DASYURIDÆ ...	... 267
K. THYLACINUS, Temm....	... 267
1. spelæus, Owen ...	... 267
L. SARCOPHILUS, F. Cuvier ...	... 267
1. laniarius (Owen) ...	... 267
M. PROTHYLACINUS, Amegh. ...	... 268
1. patagonicus, Amegh. ...	... 268
N. AMPHIPROVIVERRA, Amegh.	... 268
1. manzaniana (Amegh.) ...	... 269
O. PERATHEUTES, Amegh. ...	... 269
1. pungens, Amegh. ...	... 269
FAMILY DIDELPHYIDÆ ...	... 271
P. DIDELPHYS, Linn. ...	... 271
FAMILY TRICONODONTIDÆ	... 273

	PAGE
Q. TRICONODON, Owen	... 274
FAMILY SPALACOTHERIIDÆ	... 275
R. SPALACOTHERIUM, Owen ...	... 275
FAMILY AMPHITHERIIDÆ...	... 275
S. PHASCOLOTHERIUM, Owen ...	... 276
T. AMPHILESTES, Owen	... 276
U. AMBLOTHERIUM, Owen ...	... 276
FAMILY DROMATHERIIDÆ ...	... 277
V. DROMATHERIUM, Emmons	... 277
FOSSIL MONOTREMATA	... 279
FAMILY ORNITHORHYNCHIDÆ	... 279
W. ORNITHORHYNCHUS, Blumenb.	... 279
1. agilis, De Vis ...	... 279
FAMILY ECHIDNIDÆ	... 279
X. ECHIDNA, Cuv. ...	... 279
1. oweni, Krefft ...	... 279
ORDER MULTITUBERCULATA	... 280
FAMILY TRITYLODONTIDÆ	... 282
Y. TRITYLODON, Owen	... 282
FAMILY BOLODONTIDÆ	... 283
Z. BOLODON, Owen	... 283
FAMILY POLYMASTODONTIDÆ ...	... 283
A. POLYMASTODON, Cope... ...	... 283
FAMILY PLAGIAULACIDÆ ...	. . 284
B. PLAGIAULAX, Falconer ...	... 284

INDEX TO APPENDIX.

FAMILY MACROPODIDÆ.
DENDROLAGUS :
    5. bennettianus (De Vis, Scl., Waite) ... 287
FAMILY PHALANGERIDÆ.
PHALANGER :
    3a. breviceps (Thomas) 288

SYSTEMATIC INDEX.

PAGE

PSEUDOCHIRUS:
11. dahli (Collett) 288
FAMILY EPANORTHIDÆ 289
CŒNOLESTES 290
  1. fuliginosus (Thomas) 292
  2. obscurus (Thomas) 292
FAMILY DASYURIDÆ.
PHASCOLOGALE:
  1. cristicauda (Spencer) 293
  13. calura (Spencer) 294
  14. macdonellensis 294
SMINTHOPSIS:
  5. larapinta (Spencer) 295
  6. psammophilus (Spencer) 296
DASYUROIDES 297
  byrnei (Spencer)... 298
FAMILY DIDELPHYIDÆ.
DIDELPHYS:
  1a. koseritzi (Ihering) 299
  5a. trinitatis (Thomas) 299
  11a. soricina (Philippi) 300
  12. marmota (Oken) 300
  12a. incana (Lund) ... .: .. 301
  12b. fuscata (Thomas) 301
DROMICIOPS 302
  australis (Philippi) 302
FOSSIL FORMS 303

CORRIGENDA.

Page 63, line 17 from top, *for* toe *read* two.
,, 128, line 6 from bottom, *after* three *add* or four.
,, 128, line 12 from bottom, *for* Polyprotodon *read* Polyprotodont.
,, 150, line 9 from top, *for* five *read* three.
,, 150, line 11 from bottom, *for* carnivorous *read* herbivorous.
,, 151, line 12 from top, *for* four *read* three.
., 166, line 15 from top, *for* fine *read* five.
,, 166, line 3 from bottom, *for* inscetivorous *read* insectivorous.
,, 181, line 10 from bottom, *before* limbs *add* hind.
,, 193 line 6 from bottom, *for* three *read* four.

LIST OF PLATES.

I.—Great Grey Kangaroo	...	*Macropus giganteus.*
II.—Black-tailed Wallaby		,, *ualabatus.*
III.—Red-necked ,,		,, *ruficollis.*
IV.—Parry's ,,		,, *parryi.*
V.—Dama ,,	...	,, *eugenii.*
VI.—Brush-tailed Rock-Wallaby		*Petrogale penicillata.*
VII.—Queensland Tree-Kangaroo	...	*Dorcopsis lumholtzi.*
VIII.—Common Rat-Kangaroo	...	*Potorous tridactylus.*
IX.—Brush-tailed ,,	...	*Bettongia penicillata.*
X.—Koala	...	*Phascolarctus cinereus.*
XI.—Common Phalanger ...	...	*Trichosurus vulpecula.*
XII.—Common Ring-tailed Phalanger		*Pseudochirus peregrinus.*
XIII.—Tasmanian ,, ,,		,, *cooki.*
XIV.—Taguan Flying Phalanger	...	*Petauroides volans.*
XV.—Squirrel ,, ,,	...	*Petaurus sciureus.*
XVI.—Lesser ,, ,,	...	,, *breviceps.*
XVII.—Common Dormouse Phalanger	...	*Dromicia nana.*
XVIII.—Pigmy Flying Phalanger	..	*Acrobates pygmæus.*
XIX.—Tasmanian Wombat ...	...	*Phascolomys ursinus.*
XX.—Common Rabbit-Bandicoot	...	*Peragale lagotis.*
XXI.—Gunn's Bandicoot ...	...	*Perameles gunni.*
XXII.—Long-nosed Bandicoot	...	,, *nasuta.*
XXIII.—Short-nosed ,,	...	,, *obesula.*
XXIV.—Tasmanian Thylacine	...	*Thylacinus cynocephalus.*
XXV.—Spotted-tail Dasyure	.. ...	*Dasyurus maculatus.*
XXVI.—Common ,,	... ...	,, *viverrinus.*
XXVII.—Yellow-footed Pouched-Mouse	...	*Phascologale flavipes.*
XXVIII.—Brush-tailed ,,	...	,, *penicillata.*
XXIX.—Common ,,	...	*Sminthopsis murina.*
XXX.—Banded Ant-eater	...	*Myrmecobius fasciatus.*
XXXI.—Marsupial Mole ...	...	*Notoryctes typhlops.*
XXXII.—Common Opossum	...	*Didelphys marsupialis.*
XXXIII.—Rat-tailed ,,	...	,, *nudicaudata.*
XXXIV.—Woolly ,,	... ...	,, *lanigera.*
XXXV.—Three-striped Opossum	... ...	,, *americana.*
XXXVI.—Water ,, ...	...	*Chironectes minimus.*
XXXVII.—Duck-bill	...	*Ornithorhynchus anatinus*
XXXVIII.—Common Echidna	... ...	*Echidna aculeata.*

MARSUPIALS AND MONOTREMES.

PART I.

THE MARSUPIALS—ORDER MARSUPIALIA.

INTRODUCTION.

DIFFERING widely from all other regions of the globe as regards both its fauna and flora, the great island-continent of Australia, together with certain of the South-eastern Austro-Malayan islands, is especially characterised by being the home of the great majority of that group of lowly Mammals commonly designated Marsupials, or Pouched Mammals. Indeed, with the exception of the few species of the still more remarkable Monotremes, or Egg-laying Mammals, nearly the whole of the Mammalian fauna of Australasia consists of these Marsupials, the only other indigenous Mammals being certain Rodents and Bats, together with the native Dog, or Dingo, which may or may not have been introduced by man. All the other Orders, such as the Ungulates, or Hoofed Mammals, the Apes and Lemurs, and the Carnivores are conspicuous by their absence from the Australian landscape, where their respective places are taken by the numerous representatives of the Marsupial order, which have adapted themselves to all modes of life. We have, for instance, both terrestrial and arboreal types, while one form recently discovered passes an underground existence like the Mole. Some again are carnivorous and others herbivorous: while among the former certain kinds live on flesh, and others on insects; an equal diversity obtaining among the vegetable-feeders, some of which live on roots, others on grasses or leaves, others on fruits, and yet others on honey or the juices of flowers. We may notice, moreover, a curious kind of parallelism between certain Marsupials and the Mammals of other regions. For instance, the carnivorous Thylacine of Tasmania so closely resembles, both in form and habits, a Wolf, that by the colonists it is universally spoken of as the "Tasmanian Wolf"; while some of the small carnivorous forms are remarkably like Weasels and Civets. Among the vegetable-feeders, on the other hand, the Wombats simulate the Marmots; the Flying Phalangers are so like Flying Squirrels that, were it not for their very different front teeth, the one might readily be mistaken for the other. A still more extraordinary resemblance exists between certain small Mouse-like and Dormouse-like Marsupials and the true Mice and Dormice; this being so marked as regards general external appearance, that it may even at first sight deceive the practised naturalist. More remote is the resemblance between the arboreal Marsupial known as the Koala and a small Bear, although it is still sufficiently marked for the Australian creature to have received the name of the "Native Bear."

Having their head-quarters in Australia and New Guinea, where they form the dominant part of the Mammalian fauna, and where alone the Egg-laying Mammals are met with, the Marsupials gradually diminish in number and importance in the islands of the north-west, being largely mingled in Celebes and the neighbouring islands with types of Mammals characteristic of the Indian or Oriental region. If, however, we pass westward across the deep channel separating the island of Celebes from

Borneo, or its continuation which runs between Lombock on the east, and Bali, at the extremity of Java, on the west, we shall find all the islands lying to the westward of that line devoid of Marsupials, and possessing a Mammalian fauna closely akin to that of India. The line of the channel in question is, therefore, evidently a very important one as regards distributional zoology; and since this importance was first clearly demonstrated and explained by the great explorer and naturalist, Dr. A. R. Wallace, it is now by common consent appropriately denominated "Wallace's line." To the eastward Marsupials extend as far as New Ireland and the Solomon Islands, but they are unknown in Polynesia proper, as, indeed, they are in New Zealand, where, by the way, save for a few Bats and perhaps a Rat, there are no indigenous Mammals at all.

Such is, roughly speaking, the general distribution of the Marsupials of the Australian region. It must not, however, be supposed that the group is confined to that part of the world, although it there attains a much greater development than elsewhere. As a matter of fact, a single family of Marsupials —namely, the Opossums—is found throughout the wooded districts of Central and South America, extending southwards to the open pampa of Argentina, and represented in North America by a single species which ranges as far north as the Hudson river and Missouri. These Opossums, which, by the way, must not be confounded with the so-called Opossums (really Phalangers) of the Australian colonists, are all carnivorous, but differ widely in structure from the members of the group inhabiting Australia. That they were originally Old World forms is proved by the occurrence of their fossil remains in the middle Tertiary deposits of England and other parts of Europe, where Marsupials allied to those of Australia are now unknown. If, however, we descend deeper in the geological scale, and reach the lower portion of that great series of rocks which underlies the chalk and greensands, and is collectively termed the Jurassic or Oolitic system, we shall find that Marsupials, more or less closely allied to the most primitive carnivorous types now inhabiting Australia, were once widely spread over Europe. As these Marsupials are quite unknown after the Jurassic period, we may assume that during or soon after that time they reached Australia, where they have ever since been completely cut off from all the rest of the Old World, save New Guinea and the islands above mentioned.

Before leaving this part of our subject, we may refer somewhat more fully to the views of Dr. Wallace, as displayed in his most recent work, with regard to the former connections of Australia. After stating that the geological relations of the western Malayan islands are undoubtedly with continental Asia, he gives his reasons for considering that the very shallow seas connecting the islands of the northern area indicate that not only Java and Borneo, but likewise the Philippines, formed a south-eastern extension of the Asiatic continent at a comparatively recent date; this connection being still more clearly demonstrated by the evidence of the fauna and flora. Passing over the deep seas separating this original land-mass from New Guinea and Australia, we at once enter the area distinguished by the peculiarities of its fauna mentioned above, and which, as we have said, appears to have been isolated from the Asiatic continent since the Jurassic period. A study of New Guinea, the Moluccas, and the islands as far as Lombock in the north, and Tasmania in the south, seems further to prove that the land of Australia was formerly much more extensive that it is at present. So far as the eastern coast is concerned, this is amply demonstrated by the great barrier reef, which indicates the original limits of the land in this direction. On the same coast, but at a greater distance from the land, are several scattered islands, among which is New Caledonia.

As the latter is still subsiding, there is accordingly evidence of a still further extension of the land in this direction at an earlier date. Wallace therefore concludes that Australia must be regarded as an ancient continent of the Secondary or early part of the Tertiary period, which is now gradually decreasing in size, this diminution being indicated by the gradual subsidence of New Caledonia and certain other South Pacific islands.

Till quite recently it was likewise believed that there were no traces of Australian types of Marsupials in the Tertiary rocks of the New World. Of late years, however, there have been discovered in Patagonia a number of remains of large carnivorous Marsupials with teeth and jaws of the same general type as those of the Tasmanian Thylacine, and apparently belonging to the same group. If this eventually prove to be true, it would seem to point to there having been some kind of land connection between Australia and South America long after the period when the former country was completely sundered from the rest of the Old World; and it is noteworthy that there are certain other lines of evidence pointing to the same conclusion. Be this, however, as it may, it is certain that the Australian Marsupials, having been isolated for countless ages from the rest of the world, and thus not exposed to the competition of the higher types of Mammalian life, have flourished and developed to an extent which they could not possibly have reached in any other part of the world under existing conditions. Even there, however, their present state of development is nothing to what it was during the Pleistocene or latest geological epoch, since we find at that period evidence of the existence of giant Kangaroos and Wombats (to say nothing of certain extinct forms which have left no living kindred), by the side of which the largest existing species would appear almost dwarfs. The cause of this universal extinction (for universal it is) of the most gigantic Mammals throughout the world soon after man made his appearance on the earth, is one of those problems which has not yet received a satisfactory answer, as not even a glacial period could have made a clean sweep of the whole globe.

Having said thus much as to the general distribution of the Marsupials, we naturally turn to the consideration of what constitutes a Marsupial, or in what respects the group differs from other members of the Mammalian class. Since these animals derive their names from the presence of the well-known pouch or *marsupium* in which the young of so many species are carried, it would be not unnatural to suppose that the presence of this pouch would constitute the chief distinctive feature of the whole group. Unfortunately, however, the pouch is wanting in certain representatives of the assemblage, and we have accordingly to rely on other characteristics in order to define a Marsupial. Such a characteristic is to be found in the extremely imperfect state of development in which the young are brought forth; these being in fact little more than helpless and almost motionless lumps or sacs of flesh, which are exceedingly small in relative size to their parents, the new-born young of the Great Red Kangaroo being not larger than a man's thumb. At birth, these imperfectly-developed young are transferred to the teats of the female parent, where they hang suspended by the aid of special grasping muscles in their thick and nearly circular lips. They are, however, quite unable to suck by themselves, and the milk is consequently injected down their throats by means of a muscle which has the power of compressing the mammary gland. In all members of the group the teats are situated on the abdomen, and they are generally placed within that pouch or *marsupium*, from which the order takes its name, the only exceptions to this rule being the instances where the pouch itself is absent. Showing at first but little signs of life, the young Marsupials hang on to the nipples of their parents till they have attained a state of development permitting them to move and perform the other ordinary functions of life; they may, however, as among the Kangaroos, still make use of the pouch as a place of refuge till they attain dimensions which prohibit their entrance. It will be inferred from the above that the placenta or vascular organ connecting the blood-vessels of the fœtus with those of the parent in the higher Mammals is totally wanting among the Marsupials, whence they are often spoken of as Implacentals, in contradistinction to the Placentals, or higher Mammals. With the exception of the Thylacine, where they are rudimentary, Marsupials are further characterised by the presence of a pair of so-called marsupial, or epipubic bones, attached to the front edge of the middle of the lower part of the pelvis. It was at one time supposed that these bones bore some special relation to the pouch, but, since they are present in both sexes, it is obvious that such cannot be the case.

Among other features common to all Marsupials, it may be mentioned that the brain is relatively very small in proportion to the size of the head and body, while its external surface exhibits comparatively few foldings and convolutions; thus indicating that the brain-power and general intelligence of these creatures is of a low grade. Another indication of their low organisation is afforded by the structure of the uterus of the female, which is double throughout its length, whence the name of Didelphia has been applied to the group. Another very characteristic feature is the inward inclination or inflection of the lower border of the angle, or hinder extremity of the inferior edge, of the lower jaw.

In all cases there are two pairs of limbs, which are normal

in their development, although the hinder pair are invariably the larger of the two, and form the chief agents in progression; the greatest development of the hind limbs occurring among the Kangaroos. As a rule, there is a well-developed tail, this appendage being generally long and tapering, in some cases prehensile, and at others forming an additional support to the body. Great variation obtains in regard to the form and number of the teeth, which may be adapted either for a carnivorous, insectivorous, or herbivorous diet, although they are always divided into incisors, canines, premolars, and molars. There is, however, a marked peculiarity in regard to the succession of the teeth, which may be considered as characteristic of the order. Thus, instead of the whole, or nearly the whole of the first set of teeth in advance of the true molars or hindmost cheek-teeth, being replaced by a second set of permanent teeth, only one tooth is thus replaced, and that not invariably so. The tooth thus replaced has been hitherto generally regarded as corresponding to the last or fourth milk-molar of the higher Mammals, while the apparently replacing tooth has consequently been identified with the last or fourth premolar of the same. Recent researches have, however, tended to show that this is not a true case of replacement at all, and that the tooth which makes its appearance late in life is really a retarded premolar, which will consequently be the fourth of the full series, while the apparently replaced tooth is really the fifth. Be this as it may, the mode of succession is peculiar and unique; and it may be still convenient to speak of the replacing tooth as the fourth premolar, and the one it displaces as the fourth milk-molar. It is worthy of mention, as an admirable instance of adaptation to peculiar conditions, that during the period while the helpless young are adherent to the teats of their parents their breathing organs are specially modified in order to prevent the danger of choking by the forced injection of milk from the mammary glands. This is brought about by the upper part of the larynx, or extremity of the wind-pipe, being at this period of existence so much elongated as to project through the fauces or hinder aperture of the mouth and reach the internal nostrils. By this means a closed passage is formed from the nostrils to the lungs, and the young creature can thus breathe securely while a stream of milk is flowing down its mouth and throat into the stomach. It is noteworthy that this condition, which is merely transitory among the Marsupials, persists throughout life in the Cetaceans.

The whole of the animals presenting the foregoing characteristics constitute one Order, to which zoologists apply the name of Marsupialia. The differences by which the Marsupialia are distinguished from the higher orders of Mammalia are, however, so much greater than those by which the latter are severally differentiated from one another, that it has been further considered advisable to regard the whole of the latter as constituting one great Sub-class, while the Marsupials form a second Sub-class to themselves. For the Sub-class comprising all the higher Mammals naturalists have adopted the names of Monodelphia, Placentalia, or Eutheria; while for the one represented solely by the Marsupialia, the names of Didelphia, Implacentalia, and Metatheria have been proposed.

So different in general form and appearance from one another are many of the Marsupials—some, like the Thylacine, being carnivores, adapted for walking, while others, like the Kangaroos, are herbivorous, with limbs suited for leaping—that it might at first sight appear that the Sub-class Didelphia might well be subdivided into distinct orders, in the same manner as the Monodelphia or Eutheria are split up into the Primates, **Carnivora**, **Rodentia**, **Ungulata**, &c. It has been found, however, that structurally all the Marsupials are so intimately connected with one another that any such subdivision is impracticable; and the most that can be done in this way is to divide the Order Marsupialia into two Sub-orders, mainly distinguished from one another by the form and structure of their teeth. It is true, indeed, that the more typical representatives of these two groups also differ from one another in regard to the structure of the hind foot; but there is one family which is intermediate in this respect, and thus shows how close is the relationship of the two groups.

As already mentioned, the Marsupials have adapted themselves to almost all modes of life, some running in the ordinary manner, some progressing on the ground by long leaps, others being arboreal, and others volant, after the manner of the Flying Squirrels, while a single species has taken to a subterranean Mole-like life. It is, however, very remarkable that not a single Australian representative of the Order is aquatic in its habits, so that such an animal as a "Marsupial-Otter" does not exist in that region. The place in nature thus left vacant by the Marsupials has been seized upon by the Duck-bill and by two members of the Rodent order (otherwise so poorly represented in Australia), of which the best-known is commonly termed the Beaver-Rat (*Hydromys chrysogaster*). In South America one of the Opossums—the Water-Opossum—has, however, assumed aquatic habits, having webbed feet, and passing most of its time in the waters of rivers and streams.

Except for their skins, some of which are imported in large numbers to Europe, where they have a considerable value, the Marsupials are of little commercial importance to man. None of the herbivorous species have been domesticated as a source of food-supply, their flesh being apparently but seldom eaten at all; while all the larger carnivorous kinds are killed off as pests whenever met with, one or two having indeed been well-nigh exterminated in many districts.

THE KANGAROOS. FAMILY MACROPODIDÆ.

The Kangaroos and their allies, or *Macropodidæ*, form the first, and, in some respects the most, specialised Family of the first Sub-order of the Marsupials, known as the Diprotodontia; and before considering the characteristics of the family, it will first be necessary to mention those of the Sub-order.

The Diprotodonts (so named from the general presence of only a single pair of incisor teeth in the lower jaw) form a large group exclusively confined to Australia and the neighbouring islands, and include the whole of the herbivorous representatives of the Order. They are especially characterised by the

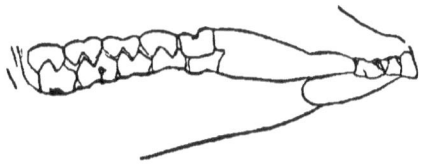

Side View of Upper and Lower Teeth of Kangaroo.

number of their incisor teeth, which never exceed three pairs above and below, and are usually three on each side of the upper and one in the lower jaw, although in the Wombats there is but a single pair in each jaw. In all cases the innermost (in some instances the only) pair of incisors are large and furnished with cutting edges; while usually in the upper, and invariably in the lower jaw the canines are either small or wanting, and never form conspicuous tusks. On the other hand, their molar teeth (four in number) have broad and squared crowns, generally carrying either a pair of transverse ridges or several blunt and rounded tubercles, and thus being in all respects adapted for a herbivorous diet.

As a family, the *Macropodidæ* are distinguished from the other two subdivisions of the Diprotodonts by the following characters: All the teeth are rooted, and there are three pairs of upper and one of lower incisors, while there may be a small canine in the upper jaw; the lower incisors being nearly horizontal, and frequently working against one another like the

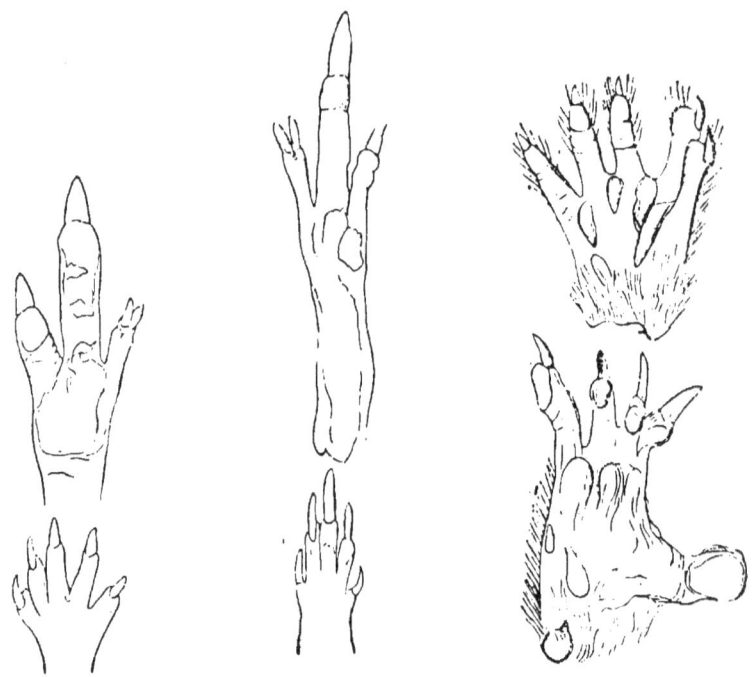

Left Hind and Fore Foot of a Kangaroo (reduced), and the corresponding but opposite Feet of a Rat-Kangaroo (*Potorous*). For comparison are added figures of the Fore and Hind Foot of a Phalanger (*Pseudochirus*).

blades of a pair of scissors. The number of cheek-teeth on each side of both the upper and lower jaws is six (although some of the earlier ones may be lost in the adult); and of these the four true molars may have either transversely ridged or tuberculated crowns,

The fore limbs, which are always provided with five toes, are generally much smaller than the hinder ones; the latter having usually only four toes, of which the one corresponding with the fourth in the human foot is much larger than either of the other three, and forms the sole axis of support. Of the remaining toes, the two innermost, or those representing the second and third human toes, are very short and small, and enclosed in a common skin; this type of foot being accordingly termed *syndactylous*. The outermost, or fifth toe, is equally small and unimportant, being even still shorter than the second and third. In consequence of this great development and peculiar structure of the hind foot, the more typical members of the Family progress on the ground by a series of enormous leaps; the body being also supported by the long, heavy, and cylindrical tail, of which, among the smaller forms, the tip may be prehensile. There are, however, certain aberrant members of the Family which are arboreal in their habits, although all agree in subsisting exclusively on grass or leaves. In order to aid in the digestion of such substances, the stomach is complicated by sacculations. The female is furnished with a large pouch for the reception of her young, with its aperture directed forwards.

The Kangaroos, which are spread over both Australia and New Guinea, include the largest of the living Diprotodonts, and indeed of all the Marsupials. While the majority of the species inhabit open grassy plains, others frequent scrub-jungle and rocks, while, as already said, a few species—and more especially in New Guinea—are dwellers in trees. Then, again, whereas most of the larger species are met with in large droves, termed by the colonists "mobs," the smaller kinds usually associate in small parties or pairs.

The existing members of the Family are arranged in twelve generic groups, which are brigaded under three distinct Sub-families, the third containing only a single genus differing very markedly from all the others and forming a kind of connecting-link between the present Family and the following one of the Phalangers.

THE KANGAROOS AND WALLABIES. GENUS MACROPUS.

Macropus, Shaw, Nat. Miscell., vol. i., text to pl. xxxiii. (1790).

The sub-family *Macropodinæ*, of which this genus is the first representative, includes forms of variable, although generally large, size, in which the claws of the fore feet are sub-equal and not disproportionately large, and the ears are generally long and pointed. In the teeth, the fourth upper premolar has its axis placed in the line of the other cheek-teeth, or but slightly twisted outwardly; the molars are transversely ridged, and increase in size from the front towards the back of the series; the canine is generally minute or absent; and the innermost upper incisor is but slightly longer than the other two.

The members of the large genus *Macropus*, which vary in size from that of a rabbit to that of a man, have a naked nose, well-developed ears, and the fur on the nape of the neck almost invariably directed downwards. The hind limbs are very much longer and stronger than the front pair, with their central claws markedly exceeding the terminal pad on the sole of the foot in length. The thick and tapering tail is evenly haired and not bushy; and the teats are four in number.

All the species of the genus are terrestrial and saltatorial in their habits; their range including Australia and the eastern Austro-Malayan islands. They may be divided into three groups, the first of which is represented by the large Kangaroos proper, while the second includes the considerably smaller animals which may be termed Larger Wallabies; the last group consisting of the Smaller Wallabies. The species of these several

groups are all closely allied to one another; and although the groups are readily distinguished, yet their distinctive characters are too trivial to be regarded as of generic value.

I. THE GREAT GREY KANGAROO. MACROPUS GIGANTEUS.

Yerboa gigantea, Zimm., Spec. Zool. Geogr., p. 526 (1777).
Macropus giganteus, Shaw, Nat. Miscell., vol. i., pl. xxxiii. (1790); Thomas, Cat. Marsup. Brit. Mus., p. 15 (1888).

(*Plate I.*)

Characters.—Size large; nose hairy between the nostrils; central hind claw long; colour greyish-brown, with the underparts and limbs nearly white. Form comparatively slender and graceful; fur short, close, and rather woolly, its direction on the fore part of the body not as constant as in the species with coarser hair. Face coloured like back, with a rather darker "whisker-mark" on sides of nose; tail brown, gradually darkening to the black tip. Length of head and body about 60 inches; tail about 36 inches.

Distribution.—The whole of Australia, with the exception of the extreme north; replaced in Tasmania by the variety *M. fuliginosus*, which is characterised by the longer, darker, and coarser hair, of which the general colour is dull smoky grey, without any tinge of fawn, the tail being grizzled grey, with its terminal fourth deep black. A second race (*M. melanops*), known as the Black-faced Kangaroo, and found both in Eastern and South-eastern Australia, is characterised by its smaller size, slighter build, and darker coloration. The general hue is dark brown, with the face darker than the back, and a brown patch across the muzzle connecting the two "whisker-marks"; while the limbs are not paler than the body.

Habits.—This species—the Koora of the natives, and the "Old Man" or "Forester" of the colonists—is the commonest of the larger Australian species, and (with the exception of *Macropus brunii*) was the one first discovered by Europeans, having been met during Cook's first voyage in July, 1770. Found in large companies or "mobs," which used to average from fifty to sixty individuals each, although sometimes containing as many as a hundred and fifty head, the Great Grey Kangaroo frequents either open plains or timbered jungle; and its general habits may be taken as typical of all the larger species of the genus. When at rest in the upright position, in which it stands from four to five feet in height, the Kangaroo supports itself upon its hind feet and haunches, with the tail, which cannot be bent, stretched out at full length upon the ground. When leaping, the short fore limbs are held close into the chest; and its leaps are of such length that fallen trees and low fences are taken in its stride, the length of a single leap being probably not far short of thirty feet. In leaping, considerable assistance is afforded by the powerful tail, of which the regular blows on the ground may be heard a considerable time before the animal itself comes into sight when it is running in thick covert. In disposition Kangaroos are extremely timid and wary; their senses of smell and hearing being very highly developed. The different "mobs" are said to keep entirely separate; and after the breeding-season the old males frequently separate from the rest of the herd to dwell apart. Their favourite haunts are open rising grounds, more or less covered with timber; and they feed both at early morning and in the evening, as well as probably in the night. Tender shoots of shrubs and herbage appear to form their food nearly or quite as much as the various grasses, and in cultivated districts they do much harm to growing crops. During the daytime in winter these animals love to lie on dry sandy rises, basking in the sun, where they may be seen by the observer who approaches them with due caution stretched out in all conceivable postures, at times lazily scratching themselves with one paw. The fierce heats of summer compel them, however, to seek protection from the sun's rays in the cool and thickly-wooded gullies, from which they only venture forth with the falling shades of evening.

The pairing-season generally takes place about January or February, and the young are born soon after. Although the pouch of an old doe usually shows three teats which have been suckled, only a single offspring is generally produced at a birth, and there is but one birth in a year. The young helpless Kangaroo is believed to remain attached to the teat in the pouch for about sixty days, not relinquishing its hold till fully developed. For some time after its removal from the teat, however, it spends the greater part of its existence in the pouch, not leaving it for long till able to run by the side of the dam. Even then, however, the "Running Joey," as the young is then called by the colonists, seeks refuge in the pouch at the first sign of danger; and it is remarkable with what agility the female will pick up and consign to her pouch the young Kangaroo, even when both are going at their best speed. When ensconced in the pouch the "Joey" always has its face protruding. Although the female will do her utmost to protect her young, when hard pressed she generally ejects it from the pouch in order to try and save her own life. By September the young are always able to run well, although they may occasionally be seen in full activity by the middle of July. About the latter part of December—the Australian midsummer—the young generally finally leave their mothers, whose pouches they are then too big to enter, and retire to form herds by themselves.

For the greater part of the year the two sexes of the Great Grey Kangaroo live together in the most perfect amity, but during the breeding-season the old males frequently engage in the most terrific conflicts, the marks of which are borne by many of them till their dying day.

For many years a merciless war of extermination was waged by the colonists against this and the other species of the genus, in consequence of which they were greatly reduced in number. Of late years they are, however, said to have increased considerably in many districts; this being partly attributed to the destruction of the Dingoes, and partly to an appreciation of the value of the Kangaroos themselves, the skins and fur of which now form an important article of export. The Grey Kangaroo is hunted with powerful dogs somewhat akin to the Scotch deerhound; and while difficult enough to seize when running on broken ground on account of its enormous flying leaps, when brought to bay it is a dangerous antagonist, seizing the dogs with its fore paws and dashing them to pieces, or ripping them open with a blow from the great conical claw of the central toe of the hind foot. Owing to its muscular strength, the tail is enabled to support the whole weight of the body during the moment that a blow is delivered by the hind legs.

As already stated, the Great Grey Kangaroo was first discovered in 1770 by Captain Cook's expedition, and the following sentence is taken from the original account of the discovery: "On Friday, June 22nd, a party who were engaged in shooting Pigeons for the use of the sick of the ship, saw an animal which they described to be as long as a Greyhound, of a slender make, of a mouse-colour, and extremely swift." The same kind of animal was soon after seen by several of Cook's party, among them, by Cook himself and Banks; while eventually they had an opportunity of examining a specimen shot by one of the members of the expedition.

Regarding the general appearance of the Kangaroo when in motion, Waterhouse writes as follows: "Like other animals whose forms differ considerably from those with which we are familiar, the Kangaroo, when first beheld, does not strike us as having agreeable proportions, and its movements appear awkward, especially when the animal is browsing, at which time it rests upon its four legs. Requiring then to move but short distances, the body is outstretched, and the hinder parts, assisted by the tail, are suddenly brought close to the anterior extremities, and this movement is repeated so long as the animal continues to graze; but when it wishes to reach a distant spot, the fore legs are removed from the ground, and it attains its end by a succession of bounds, and with an ease which at once removes the impression of awkwardness."

About one hundred thousand skins of this and other large Kangaroos are stated to be annually sold in London; the price of large skins, which may weigh as much as a couple of pounds, reaching to three shillings, or even more, per pound. As a rule, the larger skins are dressed for leather, being found especially suitable for japanning; but the skins of younger beasts, in which the fur is longer, are used for rugs, coats, and linings.

By most of those who have tried it, the dark red flesh of this and the other large species of Kangaroos is said to be dry and insipid, and not to be compared either for nourishment and juiciness with mutton, or for flavour with venison. The thick tail is, however, stated to afford a most excellent and nourishing soup.

When hard pressed, the Great Grey Kangaroo, like others of his kindred, will take to the water, in which he swims rapidly and strongly; and there is an instance on record when one of these animals swam in the sea for a distance of upwards of two miles against a head-wind and current.

II. THE ANTILOPINE KANGAROO. MACROPUS ANTILOPINUS.

Osphranter antilopinus, Gould, Proc. Zool. Soc., 1841, p. 80.

Halmaturus antilopinus, Schinz, Synops. Mamm., vol. i., p. 564 (1844).
Macropus antilopinus, Waterhouse, Nat. Hist. Mamm., vol. i., p. 95 (1846); Thomas, Cat. Marsup. Brit. Mus., p. 21 (1888).

Characters.—Size large; form stout and heavy. Fur very short, coarse, and straight, without under-fur. Nose entirely naked. General colour rich rufous, without face-markings, and becoming whitish on the under-parts. Fore and hind feet rufous brown, shading into black on the toes; tail rather darker at the tip, but otherwise coloured as the body. Ears short; feet rather short, with the central hind claw very short. Length of head and body about 56 inches; tail about 36 inches. *Female* smaller, and less brightly coloured, with the general hue of the fur greyish-fawn, in place of rufous. Skull with the region of the nose enormously inflated.

Distribution.—The whole of the northern territory of Australia.

Habits.—Nothing seems to have been ascertained as to the habits of this magnificent species, which is very rare in European collections; no specimens having been received between the date in which the species was originally described by Mr. Gould and the year 1888.

III. THE WALLAROO. MACROPUS ROBUSTUS.

Macropus (Petrogale) robustus, Gould, Proc. Zool. Soc., 1840, p. 92.
Petrogale robusta, Gray, in Grey's Australia, Appendix, vol. ii., p. 403 (1841).
Macropus robustus, Waterhouse, Nat. Hist. Mamm., vol. i., p. 100 (1846); Thomas, Cat. Marsup. Brit. Mus., p. 22 (1888).

Size large; form stout and heavy. Fur of medium length, rather thick and coarse. Nose naked. General colour dark smoky brown, lighter on the under-parts; region of nose and back of ears nearly black; lips, inside and base of ears pale grey; limbs and tail very dark brown, gradually passing into black at their extremities. Central hind toe very short. Length of head and body about 60 inches; tail about 36 inches.

Distribution.—The mountain ranges of Queensland, New South Wales, Victoria, and South Australia.

Habits.—As implied by its name of Rock-Kangaroo, this species, which is fully equal in size to the Great Grey Kangaroo of the plains, is an inhabitant of bare rocky ranges, where it is generally found in parties of from four to six. In disposition it is much more savage than the Grey Kangaroo, biting savagely when attacked, and at the same time making effective use of its extremely powerful fore limbs. It is even said that when extremely hard pressed, this Kangaroo will rush on its foes and hurl them by its impact down the steep ledges among which it is hunted.

Gould, who was somewhat indisposed to admit the social habits of the Great Grey Kangaroo, writes that the present species "may be regarded as a gregarious animal, four, six, and even more individuals being frequently seen in company. On one of the mountains near Turi, to the eastward of the Liverpool Plains, it was very numerous; and from the nature of this and the other localities in which I observed it, must possess the power of existing for long periods without water, that element being rarely met with in such situations. The summits of the hills to which this species resorts soon become intersected by numerous roads and well-trodden tracks, caused by its repeatedly traversing from one part to the other; its food consists of grasses, and the shoots and leaves of the low scrubby trees which clothe the hills it frequents."

The Kangaroo described under the title of *Macropus erubescens* appears to be nothing more than a somewhat rufous variety of the present species.

IV. ISABELLINE KANGAROO. MACROPUS ISABELLINUS.

Osphranter isabellinus, Gould, Proc. Zool. Soc., 1841, p. 81.
Macropus isabellinus, Waterhouse, Nat. Hist. Mamm., vol. i., p. 99 (1846); Thomas, Cat. Marsup. Brit. Mus, p. 25 (1888).

Character.—Size large; fur of medium length, very soft and fine, but not woolly. General colour rich foxy red, becoming white on the under-parts and limbs; front of neck pure white, sharply defined from the rufous nape by a ridge of opposed hairs; tail rufous grey. Size about the same as that of the next species.

Distribution.—North-western Australia and the adjacent islands.

In 1888 this fine species was only known by a single skin in the collection of the British Museum.

V. GREAT RED KANGAROO. MACROPUS RUFUS.

Kangurus rufus, Desmarest, Mamm. (Suppl.), vol. ii., p. 541 (1822).
Kangurus laniger, Gaimard, Bull. Soc. Philom., 1823, p. 138.
Macropus lanigerus, Gray, in Griffith's Animal Kingdom, p. 226 (1827)
Macropus rufus, Bennett, Cat. Nat. Hist. Austral. Mus., p. 6 (1837); Thomas, Cat. Marsup. Brit. Mus., p. 25 (1828).

Characters.—Size very large; form robust, rather slender in the female; fur of back and sides short, close, and woolly,

composed almost entirely of under-fur; its direction, more
especially on the head, variable; general colour of upper-parts
brilliant rufous in the male and bluish slaty-grey in the female;
under-parts white or pale grey, with fur coarse and straight; a
black "whisker-mark" on the face, with a whitish blotch below;
ears grey or brown externally and whitish inside; toes black;
tail grey; central hind claw short. Length of head and body
about 65 inches; of tail about 42 inches.

Distribution.—Eastern, South-eastern, and South Australia.

Habits.—This, the largest member of the whole group, is a common species on the plains of many parts of Eastern and Southern Australia, where it is often spoken of as the Woolly Kangaroo. Although the Great Grey and the Red Kangaroo may sometimes be found inhabiting the same districts, it appears from the observations of Gould that they more commonly frequent localities of a different description, the former preferring grassy valleys and situations where the dark soil is covered with brushwood, while the latter generally selects spots where the hard red stony ridges are clothed with box, or open plains, where it may bask in the rays of the sun. The two specimens obtained by the writer mentioned were each procured by a single dog; one being held at bay until the hunting party came up and despatched it, after a fearful resistance.

Till some thirty years ago this large and handsome species was, strangely enough, almost or quite unknown to the Australian colonists. About that date Krefft states that it had already become rare on the left bank of the Murray river, but that it was still found in considerable numbers in New South Wales and South Australia.

The same writer states that, like the Grey Kangaroo, this species feeds in flocks, "and, when disturbed, the old males cover the retreat of the fleet females, who are off first, so that specimens of the latter sex are rare, the dogs generally stopping the progress of the rear-guard of 'old men.' In wet weather, when the chalky top-soil of the 'malley scrub' is softened, these Kangaroos are easily captured ; they sink deep into the ground, and any black-fellow's cur, trained to such work, will stick to the tail of the Kangaroo until his master is able to come up and crack its skull or run a spear through it."

VI. OWEN'S KANGAROO. MACROPUS MAGNUS.

Macropus (*Boriogale*) *magnus*, Owen, Phil. Trans., 1874, p. 247.

Macropus magnus, Thomas, Cat. Marsup. Brit. Mus., p. 27 (1888).

Known only by a single skull in the British Museum from Northern Central Australia, distinguished from that of the preceding species by the small size or absence of the ledges on the front of the molar teeth, as well as by the longer foramina on the palate. There is a possibility that this Kangaroo may eventually prove to be identical with the Isabelline Kangaroo, of which only a single skin was known when the British Museum "Catalogue of Marsupials" was written.

VII. BLACK-TAILED WALLABY. MACROPUS UALABATUS.

Kangurus ualabatus, Less. and Garn., Voy. Coquille, Zool., vol. i., p. 161 (1826).

Macropus ualabatus, Lesson, Man., Mamm., p. 227 (1827); Thomas, Cat. Marsup. Brit. Mus., p. 30 (1888).

Halmaturus ualabatus, Gray, in Grey's Australia, Appendix, vol. ii., p. 402 (1841).

(*Plate II.*)

With this species we leave the true Kangaroos and come to the smaller animals forming the group of the Large Wallabies.

PLATE II

BLACK-TAILED WALLABY.

The species of this group, which closely resemble one another both in size and form, exhibit a more brilliant type of coloration than the sombre-hued Kangaroos; this difference being especially noticeable in the face. As indicated by the relatively large size of their hind feet, their leaping powers are considerably greater than those of the smaller Wallabies. As a rule, the palate of the skull is less completely ossified than in the Kangaroos, and the outermost upper incisor tooth is always furnished with a single well-defined external notch, near or at the centre of the crown. The molar teeth differ from those of the Kangaroos in having an external longitudinal bridge connecting the anterior ledge with the first transverse ridge; while the median bridge which usually joins the same ledge to the first transverse ridge in the former is small or wanting.

Characters.—The following are the distinctive external characters of the Black-tailed or Common Scrub-Wallaby: Size medium; form rather stout; fur long, thick, and rather coarse; general colour dark rufous grey, the rufous predominating behind; crown of head, base of ear, outside of elbow, chin, chest, and under-parts pale rufous or yellow, but the extent of the light-coloured area variable; face-markings indistinct; ears short, and coloured like the crown of the head; a dark mark behind the elbow; feet brown, becoming black on the toes; tail black. Length of head and body about 33 inches; of tail about 26 inches.

Distribution.—Coast regions of New South Wales and Victoria.

Variety.—The variety known as the Queensland Scrub-Wallaby (*M. apicalis*), from North-eastern Queensland, differs by its shorter and coarser fur, and more sharply-defined markings. The brown mark on the side of the face is continued through the eye nearly to the ear, and is sharply separated from the white "whisker-mark"; while the rufous of the lower back is richer and brighter, and the tail generally tipped with white.

Habits.—This Wallaby, like the other species of the group, is found in small parties in thick scrub-jungle, and hence it and its kindred are frequently spoken of as Brush-Kangaroos. They generally keep either deep in the scrub itself or at its edges, where they can be easily shot in the open runs. Gould says: "This well-marked species inhabits, with but few exceptions, all the thick bushes of New South Wales, especially such as are wet or humid. I hunted it successfully at Illawarra, on the small islands at the mouth of the Hunter, and on the Liverpool ranges. In the former localities it was frequently found in the wettest places, either among the high grass, and other dense vegetation, or among the thick mangroves, whose roots are washed by each succeeding tide. The islands at the mouth of the Hunter, particularly Mosquito and Ash Islands, are not unfrequently flooded to a great extent; yet it leaps through the shallow parts with apparent enjoyment, and even crosses the river from one island to another." Enormous numbers of this Wallaby are killed for the sake of their skins, which are extensively used in Canada for coats, while the larger ones are manufactured into leather. The number of skins annually imported into England ranges from about 10,000 to 20,000; the value of each skin usually varying from sixpence to one-and-ninepence, although over three shillings has been obtained. An even larger number of skins of the Red-necked Wallaby are sold in London, although their value is less than those of the present species.

VIII. RED-NECKED WALLABY. MACROPUS RUFICOLLIS.

Kangurus ruficollis, Desmarest, Nouv. Dict. d'Hist. Nat., vol. xvii., p. 37 (1817).

RED-NECKED WALLABY – Tasmanian Variety

Kangurus rufogriseus, Desmarest, *op. cit.*, p. 36.
Halmaturus ruficollis, Goldfuss, Isis, 1819, p. 267.
Macropus ruficollis, Lesson, Man., Mamm., p. 226 (1827); Thomas, Cat. Marsup. Brit. Mus., p. 33 (1888).
Macropus rufogriseus, Lesson, *loc. cit.*

(*Plate III.*)

Characters.—Size medium; form slender; nose naked; general colour of upper-parts greyish-fawn, with the back of the neck and rump bright rufous; under-parts white or greyish-white; face-markings inconspicuous; sometimes an indistinct whitish mark on the hip; feet grey, shading into black on the toes; tail grey above and white below, with an inconspicuous black tip. Length of head and body about 42 inches; that of tail about 30 inches.

Distribution.—Southern Queensland, New South Wales, and Victoria.

Variety.—Bennett's Wallaby (var. *M. bennetti*), from Tasmania, has longer and thicker fur and a more sombre tone of coloration than the typical form. Nape and rump dull rufous brown; back and ears nearly black; face-markings almost invisible; under-parts dirty greyish-white; tail darker grey. This variety has often been regarded as a distinct species, although it is obviously only a local race modified by the peculiar climate of Tasmania; it is this race which is represented in our plate.

The Red-necked Wallaby is the largest representative of the group.

Writing of the Tasmanian race, which is generally termed "Brush-Kangaroo" by the colonists, Gould states that it is extremely abundant. "Its flesh is generally eaten and highly esteemed, and its skin forms a considerable article of commerce, being largely imported from Van Diemen's Land into England for the manufacture of boots and shoes, besides being extensively used for the same purpose in the colony. It is universally dispersed over Van Diemen's Land, whose dense and humid forests afford it a retreat so secure as to preclude all chance of its extermination for centuries to come, although many thousands are killed annually. Advertisements may frequently be seen in Hobart Town newspapers stating that three thousand skins are immediately wanted, and they are quickly supplied by the settlers, servants, and shepherds at the out-stations. They are either captured by dogs or by snares set in their runs; the skins are generally taken off on the spot, and are afterwards stretched on the ground to dry; they are then sold for fourpence or sixpence each, to persons who visit the stock-stations of the interior for the purpose of collecting them, and who retail them again in Hobart Town, or Launceston, to the advertiser or others, for colonial consumption or for exportation." The average number of skins of this and allied species of Wallabies now imported into London is estimated at from 20,000 to 30,000; their value ranging from fourpence to a shilling each.

A large number of the Tasmanian form of this Wallaby were formerly kept at Knowsley, the seat of the Earls of Derby. Upon visiting the spot where they were enclosed, Waterhouse writes that: "I could almost have fancied myself in Australia; the heads of several of these Kangaroos suddenly made their appearance from amongst a quantity of heath, and upon my approach the animals sprang forth, and with a few vigorous bounds were soon out of reach." Eventually these Wallabies were turned loose in the park, but of their final fate history is silent. When at rest, this Wallaby curves its tail between the hind legs under the body; the hind limbs being thrust straight forward, and the front pair resting on the ground.

IX. GREY'S WALLABY. MACROPUS GREYI.

Halmaturus greyi, Gray, List Mamm., Brit. Mus., p. 90 (1843).
Macropus (Halmaturus) greyi, Waterhouse, Nat. Hist. Mamm., vol. i., p. 122 (1846).
Macropus greyi, Thomas, Cat. Marsup. Brit. Mus., p. 36 (1888).

Characters.—Size medium; form slender and delicate; general colour of upper-parts greyish-fawn on the back, but more rufous on the nape and back of head; under-parts pale grey, tinged with rufous; ears rufous behind, with blackish edges; face-markings distinct, a black band bordering the naked region of the nose, and a black "whisker-mark" reaching from the nose to the eye, bordered inferiorly by a white cheek-stripe, extending nearly to the ear; an indistinct light hip-stripe; limbs white or yellowish, becoming suddenly dark on the toes; tail pale grey, becoming lighter towards the tip, and with indistinct crests of hair on both upper and lower surfaces in its terminal half; central hind claw very long and slender. Length of head and body about 32 inches; of tail 29 inches.

Distribution.—South-eastern and South Australia.

Although allied in external characters to the Red-necked Wallaby, it would seem from the structure of the skull and the form and feebleness of the upper incisor teeth that this species is really more nearly related to the small Wallabies, of which it may be an overgrown member which has assumed the external characters of the larger species.

X. BLACK-STRIPED WALLABY. MACROPUS DORSALIS.

Halmaturus dorsalis, Gray, Mag. Nat. Hist., vol. i., p. 583 (1837).
Macropus (Halmaturus) dorsalis, Waterhouse, Nat. Hist. Mamm., vol. i., p. 152 (1846).

Macropus dorsalis, Waterhouse, Cat. Mamm. Mus. Zool. Soc., p. 67 (1838); Thomas, Cat. Marsup. Brit. Mus., p. 37 (1888).

Characters.—Size medium; form light and delicate; nose entirely naked; general colour above grey, becoming rufous on the fore-quarters; under-parts white or greyish-white; a narrow black line from the back of the head to the middle of the back; face-markings nearly obsolete; upper lip white; a white spot at the base of the outer edge of the ear; back of ear rufous, darkening towards the tip; a distinct white hip-stripe; fore legs rufous; hind legs grey; toes becoming black towards the tips; tail grey, with the extreme tip black; claw of the central hind toe relatively short. Length of head and body about 32 inches; that of tail 24 inches.

Distribution.—Interior of Queensland and New South Wales.

XI. PARRY'S WALLABY. MACROPUS PARRYI.

Macropus parryi, Bennett, Proc. Zool. Soc., 1834, p. 151; Thomas, Cat. Marsup. Brit. Mus., p. 39 (1888).
Halmaturus parryi, Gray, Mag. Nat. Hist., vol. i., p. 583 (1837).

(*Plate IV.*)

Characters.—Size medium; form slender and graceful; fur soft and almost woolly; general colour of upper-parts clear grey, with a bluish tinge; chin, chest, under-parts, and inner sides of limbs greyish-white; face-markings distinct; two dark "whisker-marks"; cheek-stripe pure white, extending backwards to beneath the eye; a white stripe in the middle of the upper half of the neck, bordered on each side by a darker line; ears long, with the basal half and extreme tip of the outer side brown, the rest, like the inside, white; limbs, with the toes, nearly or quite black; tail very long, pale grey, with an inconspicuous

PARRY'S WALLABY.

black or grey crest below the tip. Length of head and body about 37 inches; that of tail 32 inches.

Distribution.—Mountain-ranges of New South Wales and Queensland.

Habits.—Especially characterised by its slender build and very long tail, this appears to be a common species in the regions it frequents. The first known specimen was brought home by Sir E. Parry, and was obtained near Port Stephens. It was captured by some natives, having been thrown out of its mother's pouch when the latter was hunted. At that time it was somewhat less than a rabbit in size, and until its embarkation for England was allowed to run at liberty. It lived in the kitchen of its owner, and ran about the house and grounds like a dog, going out every night after dark into the bush to feed, and returning to its friend the cook about two o'clock in the morning. Besides the food which it obtained during these nocturnal excursions, the creature would eat meat, bread, and vegetables, and, indeed, almost everything that was offered to it. It expressed its anger, when very closely approached by other persons than the cook, by a sort of half-grunting, half-hissing, very discordant sound, which appeared to come from the throat, without altering the expression of its countenance. In the daytime it would occasionally, but not often, venture out to a considerable distance from home, in which case it would sometimes be chased home by strange dogs. From these, however, it had no difficulty in escaping, through its extreme swiftness; and it was curious to see it bounding up a hill and over the garden fence, until it placed itself under the protection of the dogs of the house, and more especially two Newfoundlands, which never failed to sally forth and repel its assailants. Such is in substance the account of his pet given by Sir Edward Parry.

XII. BLACK-GLOVED WALLABY. MACROPUS IRMA.

Halmaturus irma, Jourdain, Ann. Sci. Nat., vol. viii., p. 371 (1837).

Macropus (*Halmaturus*) *manicatus*, Gould, Proc. Zool. Soc., 1840, p. 127.

Macropus manicatus, Waterhouse, Jardine's Nat. Libr., Mamm., vol. xi., p. 223 (1841).

Macropus irma, Waterhouse, *op. cit.*, p. 222; Thomas, Cat. Marsup. Brit. Mus., p. 40 (1888).

Characters.—Size small; form slender and graceful; nose partly hairy between the nostrils; fur thick and soft. General colour of upper-parts bluish-grey; under-parts grey, tinged with yellow or rufous; face-markings well defined; two dark "whisker-marks"; cheek-stripe yellow, extending backwards to the ear; ears long, their external surfaces and the crown of the head black; inside of ear yellow, with a conspicuous black tip; a dark stripe on the back of the neck; an inconspicuous pale hip-stripe; fore legs and outside of hind legs grey; part of fore legs, as well as both feet, bright yellow; toes black; tail, with the sides and basal fourth, grey; terminal three-fourths, with a well-defined crest of stiff black hairs both above and below; extreme tip occasionally white. Length of head and body about 31 inches; that of tail 29 inches.

Distribution.—Southern portion of Western Australia, where this beautiful Wallaby, so easily recognised by its well-defined markings and characteristic double-crested tail, is the only representative of the group.

XIII. AGILE WALLABY. MACROPUS AGILIS.

Halmaturus agilis, Gould, Proc. Zool. Soc., 1841, p. 81.

Macropus agilis, Giebel, Odontographie, p. 43 (1855); Thomas, Cat. Marsup. Brit. Mus., p. 42 (1888).

Macropus papuanus, Peters and Doria, Ann. Mus., Genov., vol. vii., p. 544 (1875).
Macropus papuensis, Sclater, Proc. Zool. Soc. 1875, p. 532.
Halmaturus crassipes, Ramsay, Proc. Linn. Soc. N. South Wales, vol. i., p. 162 (1876).
Macropus crassipes, Ramsay, *op. cit.*, p. 395.
Halmaturus jardinii, De Vis, Proc. Roy. Soc. Queensland, vol. i., p. 109 (1884).

Characters.—Size medium ; form stouter and heavier than in other members of the group ; nose partly hairy between the nostrils ; fur short and coarse. General colour of upper-parts dark grizzled sandy ; flanks paler than back ; under-parts white or greyish-white ; face-markings inconspicuous ; an ill-defined dark stripe down the neck ; ears very short, the base and inside yellowish or white ; outer surface dark sandy, tipped and margined in front with black ; a dark brown stripe from the back of the nape to behind the elbow ; a well-marked white hip-stripe ; legs white or light sandy grey ; basal third of the long tail sandy, the remainder whitish, with the exception of the tip, which is blackish. Length of head and body about 37 inches ; of tail 34 inches.

Distribution.—South-eastern New Guinea, North Queensland, and Northern Territory of South Australia.

This well-marked species may be at once distinguished from its allies by its short ears, long tail, and generally uniform coloration, the markings being too inconspicuous to attract attention in a superficial view. It is the only one of the Large Wallabies found in Papua, and is the last representative of the group.

Habits.—Little seems to be recorded of the habits of either this or the last species. The latter is, however, stated to be abundant in West Australia, where it generally inhabits scrub-jungle, although occasionally seen on the open plains.

XIV. CAPE YORK WALLABY. MACROPUS COXENI.

Halmaturus coxeni, Gray, Proc. Zool. Soc. 1866, p. 220.
Halmaturus gazella, De Vis, Proc. Roy. Soc. Queensland, vol. i., p. 110 (1884).
Macropus coxeni, Thomas, Cat. Marsup. Brit. Mus., p. 44 (1888).

As this is the first representative of the group of Small Wallabies, we may first notice how the members of that assemblage differ from the Large Wallabies just described. All the members of the group are characterised by their light and elegant build and small size, several of them being no larger than a rabbit. In all the muzzle is naked, and in many of them has an upward naked extension, with the hair growing downwards on each side of it. In the skull the foramina in the front of the palate are very small, while the unossified vacuities in the hinder part of the same are of great size, being usually separated from one another in the middle line merely by a narrow bar of bone. The outermost upper incisor tooth is relatively smaller than in the large Wallabies, and carries a single well-marked notch, usually placed close to the posterior end of the crown.

This group ranges much further into the tropics than do either the true Kangaroo or the large Wallabies, one species inhabiting the Aru Islands, while a second extends through New Guinea into the New Britain group.

With these preliminary observations on the group, we revert to the Cape York Wallaby.

Characters.—Size small; form light and agile; naked portion of muzzle broad to the lip, the latter little developed; fur of medium length, thick and soft, its direction on the neck variable, sometimes inclining forwards. General colour of upper-parts grizzled grey; neck rufous; under-parts white; no streak on neck; ears long, with the back grey, margined with white; occasionally a faint light hip-stripe; legs grey or rufous; feet pale brown; tail grey for its basal fourth, the remainder brown above and white beneath. Length of head and body about 25 inches; that of tail 16 inches.

Distribution.—North Queensland.

This species and *Macropus agilis* are the only true Wallabies in North Queensland, and it is not a little remarkable that both are distinguished from the other members of the group to which they respectively belong by their short fur, general sandy coloration, inconspicuous markings, and white hip-stripe. Probably this similarity in their coloration is the result of adaptation to the surroundings of the countries which these Wallabies inhabit.

XV. BRANDED WALLABY. MACROPUS STIGMATICUS.

Halmaturus stigmaticus, Gould, Proc. Zool. Soc., 1860, p. 375.

Macropus stigmaticus, Thomas, Cat. Marsup. Brit. Mus., p. 47 (1888).

Characters.—Size medium; form light and slender; naked portion of muzzle with a central upward projection, inferiorly continued to the lip; fur short, close, and rather coarse. General colour of upper-parts rufous grey, the latter tint predominating in front and the former behind; under-parts white; crown of head, cheeks, and region round the base of the ear deep rust-colour; back of ear, hinder part of head and nape of neck brown; an indistinct darker stripe down the middle of the neck; two lateral longitudinal bright rusty bands; a conspicuous yellowish hip-stripe. Fore legs rufous; hind legs brilliant rust-colour; feet grey or rufous grey, with the tips of the toes brown; tail greyish-brown above, whitish below. Length of head and body about 28 inches; of tail 15 inches.

Distribution.—North-eastern Queensland.

XVI. RED-LEGGED WALLABY. MACROPUS WILCOXI.

Halmaturus wilcoxi, McCoy, Ann. Mag. Nat. Hist., ser. 3, vol. xviii., p. 322 (1866).
Halmaturus temporalis, De Vis, Proc. Roy. Soc. Queensland, vol. i., p. 111 (1884).
Macropus wilcoxi, Thomas, Cat. Marsup. Brit. Mus., p. 48 (1888).

This Wallaby is only a smaller and less brightly coloured non-tropical representative or variety of the preceding tropical species. The hair is longer in the southern form, and the differences between the two are similar to those between the two races of *M. ualabatus*.

Distribution.—New South Wales and Southern Queensland.

XVII. ARU ISLAND WALLABY. MACROPUS BRUNII.

Didelphys brunii, Schreber, Säugeth., vol. iii., p. 551 (1778).
Halmaturus brunii, Illiger, Prodrom. Syst. Mamm., p. 80 (1811).
Macropus brunii, Cuvier, Règne Anim., vol. i., p. 183 (1817); Thomas, Cat. Marsup. Brit. Mus., p. 50 (1888).

Characters.—Size and form medium; female markedly smaller than male: upper border of naked portion of muzzle, with only a single central prominence downwards; fur short, close, and straight, with scarcely any under-fur. General colour uniform chocolate-brown; throat and under-parts white, faintly tinged with brown; a grey grizzle on the lower rump and hips; a distinct white "whisker-mark" from the mouth to beneath the eye; ears short, black behind, and the crown of the head sometimes also black; a well-marked white hip-stripe; legs and feet, together with the tail, grey or brown, more or less grizzled with white. Length of head and body of male about 30 inches; of tail about 13 inches.

Distribution.—Aru and Kei Islands.

This species has an especial interest as being the first member of the Kangaroo-family known to Europeans, specimens having been seen in the year 1711 by Bruyn living in the garden of the Dutch Governor of Batavia. These were originally described under the name of Philander or Filander; but subsequently the species became confounded with a very different, although externally similar animal, known as *Dorcopsis muelleri*, and it was not till the two were carefully compared by the late Professor Garrod that their distinctness was established. The present species may be distinguished externally from the one last named by its very much shorter head, the backward direction of the hairs on the nape of the neck, and the distinct white stripe on the hip.

Nothing definite appears to be known as to the habits of either this or the following species, although they are probably not very different from those of their allies.

XVIII. SOMBRE WALLABY. MACROPUS BROWNI.

Halmaturus browni, Ramsay, Proc. Linn. Soc. N. South Wales, vol. i., p. 307 (1877).
Macropus lugens, Alston, Proc. Zool. Soc., 1877, p. 126.
Macropus browni, Alston, *op. cit.*, p. 743; Thomas, Cat. Marsup. Brit. Mus., p. 51 (1888).
Macropus jukesi, M. gracilis et *M. tibol*, Mikl. Macl., Proc. Linn. Soc. N. South Wales, vol. ix., pp. 890, 894, and vol. x., p. 141.

Very similar to *M. brunii*, but with thicker and softer fur, in which the general hair is greyer, with a more distinct grizzle. The ears also are brown behind like the head, the under-parts yellow instead of white, and the hip-stripe has almost disappeared.

Distribution.—New Britain group, and Eastern and South-eastern New Guinea.

XIX. PADEMELON WALLABY. MACROPUS THETIDIS.

Halmaturus thetis, Lesson, Man. Mamm. p. 229 (1827).
Halmaturus thetidis, F. Cuvier, Hist. Nat., Mamm., part lvi. (1829).
Macropus (*Halmaturus*) *thetidis*, Waterhouse, Nat. Hist. Mamm., vol. i., p. 144 (1846).
Macropus thetidis, Giebel, Odontographie, p. 43 (1855); Thomas, Cat. Marsup. Brit. Mus., p. 52 (1888).

Characters.—Size small; form light and agile; naked part of nose broad to the lip, the latter little developed; fur of medium length, thick and soft, with a variable direction on the neck. General colour grizzled grey, rufous on the neck, which has no streak; under-parts white; ears long, grey on the back, margined in front with brown or black; a faint light hip-stripe sometimes present; legs grey or rufous; feet pale brown; tail with the basal fourth grey, elsewhere brown above and white beneath. Length of head and body about 25 inches; of tail 16 inches.

Distribution.—Victoria, New South Wales, and Queensland.

Habits.—When sitting up on its hind legs, the Pademelon Wallaby is about 20 inches in height. It is very common in New South Wales, where it generally frequents scrub-covered districts. Its general habits appear to be very similar to those of other Wallabies. On account of the excellence of its flesh

PLATE V

DAMA WALLABY.

which is said to be very like that of the Hare in flavour, this species is much hunted by the colonists.

XX. DAMA WALLABY. MACROPUS EUGENII.

Kangurus eugenii, Desmarest, Nouv. Dict. d'Hist. Nat., vol. xvii., p. 38 (1817).
Halmaturus eugenii, Schinz, in Cuvier's Theirreichs, vol. i., p. 888 (1821).
Macropus eugenii, Lesson, Man. Mamm., p. 227 (1827); Thomas, Cat. Marsup. Brit. Mus., p. 54 (1888).
Halmaturus derbianus, Gray, Mag. Nat. Hist., vol. i., p. 583 (1837).
Macropus derbianus, Waterhouse, Cat. Mamm. Mus. Zool. Soc., p. 67 (1838).
Halmaturus houtmanni, et *H. dama*, Gould, Proc. Zool. Soc., 1844, pp. 31, 32.
Macropus gracilis, Gould, *op. cit.*, p. 103.

(*Plate V.*)

Characters.—Size small; form light and graceful; naked portion of muzzle ending some distance from the mouth; fur rather short in specimens from the mainland, longer in those from the islands. General colour grizzled grey, rufous on the shoulders; under-parts white or greyish-white; usually an indistinct white cheek-stripe; ears dark grey, long in specimens from the mainland, shorter in island examples; an ill-defined dark streak from the back of the head to the back; shoulders, sides of neck, and fore legs rufous; feet and tail grey, with nearly black extremities. Length of head and body from about 24 inches in mainland, to 28 in island, specimens; of tail 17 inches.

Distribution.—Western Australia, and islands off the Western and Southern Coasts.

The Dama Wallaby derives its Latin name from having been supposed to have been first obtained on Eugene Island, off the West Coast; while its English title is stated to be that by which it is known to the natives of the mainland. Allied to *Macropus thetidis* of the East Coast, it may be readily distinguished from that species by the rufous fore legs and the conformation of the upper incisor teeth.

XXI. PARMA WALLABY. MACROPUS PARMA.

Macropus (Halmaturus) parma, Waterhouse, Nat. Hist. Mamm., vol. i., p. 149 (1846).

Macropus parma, Owen, Cat. Osteol. Mus. Roy. Coll. Surg., vol. i., p. 325 (1853); Thomas, Cat. Marsup. Brit. Mus., p. 57 (1888).

Characters.—Very similar to *M. eugenii*, but with the back more rufous and not contrasting with the neck; the white cheek-stripe and dark streak on the back of the neck more distinct. Front of throat pure white, contrasting with sides of neck; under-parts greyish white; ears short, rufous grey behind.

Distribution.—New South Wales

This species seems to be very rare and locally distributed.

XXII. RUFOUS-BELLIED WALLABY. MACROPUS BILLARDIERI.

Kangurus billardieri, Desmarest, Mamm., vol. ii., p. 542 (1822).

Macropus billardieri, Lesson, Man. Mamm., p. 227 (1827); Thomas, Cat. Marsup. Brit. Mus., p. 58 (1888).

Macropus rufiventer, Ogilby, Proc., Zool. Soc., 1838, p. 23.

Halmaturus billardieri, Gould, Monogr. Macropodidæ, pl. x (1841).

Characters.—Size large; form stout and heavy; fur long, thick, and soft. General colour greyish-brown, with an olive tinge, which is most marked on the head and rump; under-parts yellow, orange, or rufous, brightest posteriorly; ears very short, olive-grey behind, margined in front with black; frequently an indistinct stripe on the back of the neck, and sometimes an indistinct yellowish hip-stripe. Legs greyish-brown; feet brown; tail very short, only about two-and-a-half times the length of the head, basally orange above, terminally greyish-white beneath, elsewhere greyish-brown. Length of head and body about 26 inches; of tail 14 inches.

Distribution.—South-eastern Australia, Victoria, Tasmania, and islands of Bass Straits.

Habits.—This is the common small Wallaby of Victoria and Tasmania, where it is extremely abundant. According to Gould's account, it is essentially gregarious, hundreds of individuals generally inhabiting the same locality. It frequents gullies and the denser and moister portions of the forest, especially such as are covered with tall rank grass, through which it makes a number of well-beaten tracks. From such coverts this Wallaby but seldom emerges, never even approaching the outskirts of the forests, except at night. Consequently, in spite of its abundance, it is but seldom seen by the ordinary observer. It is easily taken by snares, placed in the form of a noose in its runs, thousands of these Wallabies being annually captured in this manner for the sake of their skins. It is likewise highly esteemed in Tasmania on account of its flesh, which is one of the best-flavoured among the Wallabies.

XXIII. SHORT-TAILED WALLABY. MACROPUS BRACHYURUS.

Kangarus brachyurus, Quoy and Gaimard, Voyage Astrolabe, Zool., vol. i., p. 114 (1830).
Macropus brachyurus, Lesson, Hist. Nat. Mamm., vol. v., p. 378 (1836); Thomas, Cat. Marsup. Brit. Mus., p. 60 (1888).

Halmaturus (*Thylogale*) *brevicaudatus*, Gray, Ann. Mag. Nat. Hist., vol. i., p. 108 (1838).

Halmaturus brachyurus, Gray, in Grey's Australia, Appendix, vol. ii., p. 403 (1841).

Characters.—Size small; form short and squat; naked portion of muzzle with a central upward projection; fur long, thick, and coarse. General colour coarsely grizzled greyish-brown; under-parts slaty-grey; ears very short and rounded, grizzled grey, and thickly haired on the back; feet brown; tail very short, about twice the length of the head, brown above, greyish-white beneath. Length of head and body about 23 inches; of tail 10 inches.

Distribution.—Western Australia.

This species has often been confounded with the last, which it resembles in general coloration and proportions, but from which it differs by its inferior dimensions, as well as by certain well-marked characters of the skull and teeth. Externally it also resembles some of the species of *Potorous*, inhabiting the same districts, but it may always be distinguished from them by its longer feet, larger and relatively shorter and thicker head, more hairy ears, and by the grey, instead of rufous-tipped under-fur.

THE ROCK-WALLABIES. GENUS PETROGALE.

Petrogalea, Gray, Mag. Nat. Hist., vol. i., p. 583 (1837).

This genus includes six small species distinguished from the true Wallabies by the following characters. Region of nose entirely naked; fur on back of neck directed downwards; central claws of hind feet very short, only slightly exceeding the pads on the sole in length; tail long, cylindrical, and thinner than in *Macropus*, thickly haired and tufted at the tip.

PLATE VI

BRUSH-TAILED ROCK-WALLABY

The Rock-Wallabies are confined to the mainland of Australia, on which they are generally distributed, but are unknown in Tasmania. Although closely allied to the true Wallabies, their habits are markedly distinct, the Rock-Wallabies frequenting rugged rocky districts instead of the open plains. In such situations they leap and climb with remarkable agility, their long bushy tails being used solely to help in balancing the body, and never being employed as an additional support. Consequently the tail has not the same strength and thickness as in their plain-haunting cousins.

1. BRUSH-TAILED ROCK-WALLABY. PETROGALE PENICILLATA.

Kangurus penicillatus, Gray, in Griffith's Animal Kingdom, vol. v., p. 204 (1827).
Macropus penicillatus, Bennett, Proc. Zool. Soc., 1835, p. 1.
Petrogale penicillata, Gray, Mag. Nat. Hist., vol. i., p. 583 (1837); Thomas, Cat. Marsup. Brit. Mus., p. 66 (1888).
Halmaturus penicillatus, Wagner, in Schreber's Säugeth. Suppl., vol. iii., p. 126 (1843).

(*Plate VI.*)

Characters.—Size large; form stout and heavy; fur long, thick, and coarse. General colour dull brown, tending to rufous on the rump; chin and chest pale grey; remainder of under-parts brown tinged with yellow, becoming yellowish-rufous posteriorly. Whisker-mark and cheek-stripe ill-defined; a black streak on the back of the head, not extending to neck. Ears short, yellow internally and on the posterior margin, elsewhere grey at base and black terminally; a black and often inconspicuous mark behind the shoulder, followed by a pale grey one. Legs brown or rufous; toes black; tail long, more or less bushy, rufous for the basal three or four inches, elsewhere black, save sometimes the extreme tip, which may be yellow. Length of head and body about 29 inches; of tail 23 inches.

Distribution.—Coast districts of Queensland, New South Wales, and Victoria.

Habits.—These Rock-Wallabies are described as being gregarious in their habits, frequenting rocky ground, where they have holes to which they retreat when pursued. When on precipitous cliffs, they ascend the rocks in groups, jumping from side to side, and alighting on such small ledges that it seems almost impossible for them to obtain foothold. During the day they remain concealed in caves and holes, from which they issue forth at evening, while on moonlight nights they may be seen abroad at all hours. During the night, according to Gould, these Wallabies will frequently leave the well-beaten tracks among the rocks for the grass beds on the crowns or at the base of the mountains, although they never stray so far from their haunts that they are unable to regain them speedily on the slightest alarm. They also have the power of easily ascending the sloping trunks of trees, upon which they leave, in some cases, regular worn tracks. Waterhouse states that an example, in the London Zoological Gardens, was in the habit of perching itself on some narrow ledge on the walls of its enclosure, upon which it would balance its body in a manner that at first sight appeared impracticable.

II. WEST AUSTRALIAN ROCK-WALLABY. PETROGALE LATERALIS.

Petrogale lateralis, Gould, Monograph. Macropodidæ, pl. xxiv. (1842); Thomas, Cat. Marsup. Brit. Mus., p. 68 (1888).
Macropus (Heteropus) lateralis, Waterhouse, Nat. Hist. Mamm., vol. i., p. 172 (1846).
Halmaturus lateralis, Wagner, in Schreber's Säugeth. Suppl., vol. v., p. 328 (1855).

Characters.—Size medium; form slender and light; fur long, soft, close, and of a rather woolly texture. General colour light grey, with the under-parts yellowish-grey; a well-defined dark whisker-mark, with a whitish or yellowish cheek-stripe below; a narrow brown or black stripe from the back of the head to the middle of the back; ears short, with the inside, base, and extreme tip of outside yellow, and the remainder brown; a prominent black or brown mark behind the elbow, followed by a white stripe down the hip; front of knee brown, and connected by a brown band with the shoulder-spot; legs and feet grey; toes black; tail grey for the basal, and black for the terminal half. Length of head and body about 24 inches; of tail 17 inches.

Distribution.—West Australia.

Habits.—Differing from the common Brush-tailed Wallaby in its smaller size, the greyer fore-quarters, the more distinct black mark on the back, the black and white markings on the flank, and the somewhat less bushy tail, this West Australian representative of the genus is only to be met with in those parts of the interior which are rocky and well supplied with caverns. In disposition it is remarkably shy, seldom venturing out during the day, and feeding at night on the open patches of grass which occur here and there among the rocks. According to Gould, it never strays more than two or three hundred yards from its retreats among the rocks; while, when alarmed, it leaps with the most surprising agility and speed from rock to rock.

III. SHORT-EARED ROCK-WALLABY. PETROGALE BRACHYOTIS.

Macropus (*Petrogale*) *brachyotis*, Gould, Proc. Zool. Soc., 1840, p. 129.

Petrogale brachyotis, Gould, Monogr. Macropodidæ, pl. vi. (1841); Thomas, Cat. Marsup. Brit. Mus., p. 69 (1888).

Halmaturus brachyotis, Schinz, Synops. Mamm., vol. i., p. 562 (1844).

Characters.—Size medium; form light and slender; fur short and thin. General colour greyish-fawn; under-parts greyish white; face-markings almost obsolete; ears very short, fawn-grey on the back, with the edges and extreme tip white; a dark brown blotch behind the elbow, followed by a whitish band; limbs pale grey; tail grey above, whitish beneath, the terminal fourth tufted inferiorly with longer dark brown hairs. Length of head and body about 22 inches; of tail 16 inches.

Distribution.—North-west coast of Australia.

IV. PLAIN-COLOURED ROCK-WALLABY. PETROGALE INORNATA.

Petrogale inornata, Gould, Proc. Zool. Soc., 1842, p. 5; Thomas, Cat. Marsup. Brit. Mus., p. 70 (1888).
Halmaturus inornatus, Schinz, Synops. Mamm., vol. i., p. 566 (1844).
Macropus inornatus, Waterhouse, Nat. Hist. Mamm., vol. i., p. 175 (1846).

An imperfectly known form, very closely allied to *P. brachyotis*, and perhaps founded upon an individual of that species with the markings unusually indistinct.

Distribution.—North coast of Australia.

Habits.—Sir George Grey, by whom this species was first discovered in the neighbourhood of Hanover Bay, writes that it is a very wild and shy animal, frequenting in the daytime the highest and most inaccessible rocks, and only coming down in the early morning and late in the evening to feed in the valleys. When disturbed in the daytime, it bounds among the roughest and most precipitous rocks, apparently with the greatest ease, and is so watchful and wary that to obtain a successful shot is by no means an easy matter. The heat of the sand-rocks among which this Wallaby is found is stated to be tremendous, sometimes attaining at midday during the summer to as much as 136 degrees.

V. LITTLE ROCK-WALLABY. PETROGALE CONCINNA.

Petrogale concinna, Gould, Proc. Zool. Soc., 1842, p. 57; Thomas, Cat. Marsup. Brit. Mus., p. 71 (1888).
Halmaturus concinnus, Schinz, Synops. Mamm., vol. i., p. 565 (1844).
Macropus (*Heteropus*) *concinnus*, Waterhouse, Nat. Hist. Mamm., vol. i., p. 177 (1846).

Characters.—Size very small; form slender; fur short, soft, and silky. General colour rich orange-rufous; under-parts white or greyish-white; face-markings obsolete. Ears very short, pale fawn on the back. No shoulder- or flank-markings; limbs greyish-fawn; tuft of tail yellowish-brown. Length of head and body of immature male 14 inches.

Distribution.—North-east Australia.

VI. YELLOW-FOOTED ROCK-WALLABY. PETROGALE XANTHOPUS.

Petrogale xanthopus, Gray, Proc. Zool. Soc., 1854, p. 294; Thomas, Cat. Marsup. Brit. Mus., p. 65 (1888).
Macropus xanthopus, Flower, Cat. Osteol. Mus. Roy. Coll. Surg., pt. ii., p. 715 (1884).

Characters.—Size large; fur long, soft, and silky. General colour grey, white beneath; a well-defined white cheek-stripe; a rich orange spot above the eye; a black streak from the back of the head to the middle of the back. Ears long, yellow behind; a triangular brown blotch behind the elbow, followed by a white stripe down the hip; top of knee brown, with a white patch on the outer side; limbs rich yellow; tips of toes brown. Tail ringed above and on the sides with dark brown and pale yellow (the brown colour gradually coalescing above posteriorly to form a dark crest), beneath uniform yellowish- or brownish-white, tip sometimes yellow. Length of head and body about 32 inches; of tail 24 inches.

Distribution.—South Australia.

This Rock-Wallaby, which is more brilliantly coloured than any other member of the family, may be at once recognised by its ringed tail. The contrasts in its coloration are too glaring to form a pleasing effect. The skull is characterised by the inflation of the muzzle and forehead, in which respect it makes an approximation to that of *Macropus antilopinus*. Some hundreds of skins are annually imported to London from Adelaide, their value ranging from one-and-fourpence each. The skins of the common Rock-Wallaby are less valuable, averaging from threepence to ninepence each, although they have been known to reach as much as one-and-threepence.

THE NAIL-TAILED WALLABIES. GENUS ONYCHOGALE.

Onychogalea, Gray, in Grey's Australia, Appendix, vol. ii., p. 402 (1841).

Nose hairy, with the exception, in some species, of the base of the septum between the nostrils; central hind claw long, narrow, compressed, and very sharp; tail long, tapering, short-haired, not bushy, more or less crested towards the tip, where it is furnished with a horny nail or spur.

The three species of Nail-tailed Wallabies, which are confined to Australia and unknown in Tasmania, form a well-marked group, distinguished not only by the peculiar horny appendage to the tail, from which they derive their name, but likewise by the form of their incisor teeth, which are small and light, and decrease evenly in size from before backwards, the middle and outer ones in the upper jaw being very slender, and sloping markedly forwards. The use of the horny spur to the tail—paralleled elsewhere among Mammals only in some individuals of the Lion—is at present quite unknown.

1. NAIL-TAILED WALLABY. ONYCHOGALE UNGUIFERA.

Macropus unguifer, Gould, Proc. Zool. Soc., 1840, p. 93.
Onychogalea unguifer, Gray, List Mamm. Brit. Mus., p. 88 (1843).
Halmaturus unguifer, Schinz, Synops. Mamm., vol. i., p. 547 (1844).
Onychogalea annulicauda, De Vis, Proc. Roy. Soc. Queensland, vol. i., p. 157 (1884).
Onychogale unguifera, Thomas, Cat. Marsup. Brit. Mus., p. 74 (1888).

Characters.—Size large; form light and slender; nose broad, partially haired; fur thick, close, and rather short. General colour rich sandy-fawn, with a darker band down the back; under-parts white; ears thinly clothed with white hairs; an indistinct white mark behind the elbow; a white hip-stripe; limbs white, with the exception of the back of the hind legs, which is fawn. Tail very long, white above, sandy-grey beneath, the terminal third with brown rings, gradually darkening posteriorly, and finally coalescing with the black tufted tip, also forming a crest on the upper surface; terminal nail large, flattened laterally, and concealed by the pencil of hairs. Length of head and body about 26 inches; of tail about the same.

Distribution.—North-western and Northern Central Australia.

This species, of which the habits appear to be quite unknown, was first made known to science through a specimen brought home by Mr. Bynoe, of H.M.S. "Beagle."

II. BRIDLED WALLABY. ONYCHOGALE FRENATA.

Macropus frænatus, Gould, Proc. Zool. Soc., 1840, p. 92.
Onychogalea frænata, Gray, List Mamm. Brit. Mus., p. 88 (1843).
Halmaturus frænatus, Schinz, Synops. Mamm., vol. i., p. 548 (1844).
Onychogale frenata, Thomas, Cat. Marsup. Brit. Mus., p. 76 (1888).

Characters.—Size small; form light and slender; nose narrow, wholly haired; fur soft and thick. General colour clear grey; chin and chest white; rest of under-parts pale grey; middle of back of neck black; ears short, greyish-brown externally, white internally; a distinct white shoulder-stripe, continued along the sides of the neck to behind the ear; sides of neck grey, tinged with rufous; an indistinct pale hip-stripe; fore legs and outer sides of hind legs and feet white. Tail of medium length, grey, with the tip black, and the terminal nail short and rounded. Length of head and body about 22 inches; of tail 18 inches.

Distribution.—Interior of Southern Queensland, New South Wales, and Victoria.

Habits.—One of the most elegant of the Kangaroo family. This species, says Gould, "inhabits all the low mountain ranges similar to those of Brezi, the elevation of which varies from one to five or six hundred feet, and which are of a sterile character—hot, dry, stony, and thinly covered with shrub-like stony trees. . . . When started from its seat, which is formed like that of a Hare, and sheltered by a tuft of grass or a small bush, it bounds away with remarkable swiftness, generally giving the best dogs a sharp run, and frequently effects its escape by gaining the thick part of the trunk or the hole of a decayed tree; and I recollect, on one occasion, that on being sharply pressed, the animal mounted the inside of the tree to an opening nearly fifteen feet from the ground, whence it leaped down before the dogs, and succeeded in reaching the hollow trunk of a fallen tree, from which it was finally taken by hand."

Krefft states that this species is the most common of all the smaller members of the Kangaroo tribe, and that although frequently seen abroad during the daytime, in captivity is more lively by night; it is gregarious in its habits. The female, which, as in the allied forms, is considerably smaller than the male, produces a single young one at a time, generally born at the beginning of May.

III. CRESCENT WALLABY. ONYCHOGALE LUNATA.

Macropus lunatus, Gould, Proc. Zool. Soc., 1840, p. 93.
Halmaturus lunatus, Schinz, Synops. Mamm., vol. i., p. 548 (1844).
Onychogale lunata, Gould, Mamm. Australia, vol. ii., pl. lv. (1849); Thomas, Cat. Marsup. Brit. Mus., p. 77 (1888).

Characters.—Size small; form very light and delicate; nose narrow, with the base of the septum naked; fur soft and woolly. General colour dark grey; under-parts whitish; ears short, brown externally, white internally; back and sides of neck rich rufous; a conspicuous white shoulder-stripe, not extending on to back of neck; two indistinct stripes on hip; limbs pale grey; toes brown; tail short, grey, with the terminal nail as in the last species. Length of head and body about 20 inches; of tail 13 inches.

Distribution.—West and South Australia.

The habits of this species are probably identical with those of the preceding.

THE HARE-WALLABIES. GENUS LAGORCHESTES.

Lagorchestes, Gould, Monogr. Macropodidæ, pl. xii. (1841).

Nose wholly or partially haired; central hind claw long and strong, not concealed by the hair; tail rather short, evenly haired throughout, and without a terminal nail. The three species of this group are confined to Australia, exclusive of Tasmania.

1. SPECTACLED HARE-WALLABY. LAGORCHESTES CONSPICILLATUS.

Lagorchestes conspicillatus, Gould, Proc. Zool. Soc., 1841, p. 82; Thomas, Cat. Marsup. Brit. Mus., p. 80 (1888).
Halmaturus conspicillatus, Schinz, Synops. Mamm., vol. i., p. 563 (1844).
Macropus (*Lagorchestes*) *conspicillatus*, Waterhouse, Nat. Hist. Mamm., vol. i., p. 85 (1846).

Characters.—Size relatively large, and form thick and heavy; nose, with lower half of septum and edges of nostrils, hairy; muzzle broad and heavy; fur long and coarse; under-fur of back uniform blackish-brown. General colour coarsely grizzled yellowish-grey; under-parts mingled white and slaty-grey; a well-defined chestnut band round the eye, not extending forwards on the side of the muzzle; ears short, less than one-third the length of the hind foot, grizzled grey on the back, with the edges and inside nearly white; two whitish lateral bands; limbs grey, tinged with rufous; tail covered above and on the sides with scattered white hairs, except at the root, where they are grey; beneath more thickly haired, and tinged with fawn. Canine teeth well-developed and functional. Length of head and body about 20 inches; of tail 17 inches. Other characters as in next species.

Distribution.—Islands off the North-west coast of Australia;

replaced on the mainland of North Australia by the under-mentioned race, which is regarded by some as a distinct species.

Variety.—Leichhardt's Hare-Wallaby (*L. leichardti.*) has longer ears and a much more brilliant coloration than the typical form. Back deep fawn; band round the eye rich rufous; under-parts and lateral bands nearly pure white.

This species, which presents a remarkable resemblance to the English Hare, may be distinguished from the next by its shorter ears, the brighter rusty red ring round the eye, and the want of the black patch behind the fore leg, as well as by the broader muzzle.

II. COMMON HARE-WALLABY. LAGORCHESTES LEPOROIDES.

Macropus leporoides, Gould, Proc. Zool. Soc., 1840, p. 93.
Lagorchestes leporoides, Gould, Monogr. Macropodidæ, pl. xii. (1841); Thomas, Cat. Marsup. Brit. Mus., p. 83 (1888).
Halmaturus leporoides, Schinz, Synops. Mamm., vol. i., p. 549 (1844).

Characters.—General form light and slender; base of septum of nostrils naked. General colour coarsely grizzled yellowish-brown; under-parts yellowish-grey; rufous band round eye extending forward on side of muzzle; back and inside of ear whitish; a black patch on the elbow; legs like body; feet finely grizzled greyish-white; tail brownish-grey above, with the sides and under-surface nearly white. Canine teeth small. Length of head and body about 20 inches; of tail 13 inches.

Distribution.—Interior of New South Wales and South Australia.

Habits.—Of the habits of this species Gould writes as follows:

"The name of Hare-Kangaroo has been given to this species as much from its similarity of form and size to the common Hare as from its similarity of habits. I usually found it solitary, and sitting alone on a well-formed seat under the stalks of a tuft of grass on the open plains. For a short distance its fleetness is beyond that of all others of its group that I have had an opportunity of coursing. Its powers of leaping are also extraordinary. While out on the plains of South Australia, I started a Hare-Kangaroo before two fleet dogs. After running to the distance of a quarter of a mile it suddenly doubled and came back to me, the dogs following close to its heels. I stood perfectly still, and the animal had arrived within twenty feet before it observed me, when, to my astonishment, instead of branching off to the right or to the left, it bounded clear over my head, and, on descending to the ground, I was able to make a successful shot, by which it was procured."

According to Krefft, this species is common in the level country between the Murray and Darling rivers; and is strictly nocturnal and solitary in its habits. During the daytime it is generally found asleep under some salt-bush or other sheltered situation. When hunted, it takes leaps of more than eight feet in height.

III. RUFOUS HARE-WALLABY. LAGORCHESTES HIRSUTUS.

Lagorchestes hirsutus, Gould, Proc. Zool. Soc., 1844, p. 32; Thomas, Cat. Marsup. Brit. Mus., p. 84 (1888).

Macropus (*Lagorchestes*) *hirsutus*, Waterhouse, Nat. Hist. Mamm., vol. i, p. 93 (1846).

Halmaturus hirsutus, Wagner, in Schreber's Säugeth. Suppl., vol. v., p. 307 (1855).

Characters.—Form nearly as in the last; nose almost entirely hairy; muzzle narrow and light; fur long and coarse; under-fur of back dark slaty, with pale or rufescent tips. General colour finely grizzled grey, becoming rich rufous on the rump; under-parts yellowish-grey; band round the eye but slightly rufous; ears long, more than one third the length of the hind foot, grizzled grey at the back, with the inside and edges white; lateral bands inconspicuous; no black patch on the elbow; fore limbs and front of hind legs and feet pale yellowish-white or grey; outer side and back of hind legs rufous; tail short-haired, dull grizzled grey. Canine teeth small. Length of head and body about 18 inches; of tail 15 inches.

Distribution.—West Australia.

THE DORCA KANGAROOS. GENUS DORCOPSIS.

Dorcopsis, Schlegel and Müller, Verhandl. Nat. Ges. Nederl. Ind., p. 130 (1839-44).

Hind limbs proportionately less elongated than in *Macropus*; naked portion of nose large, broad, and wholly devoid of hair; head long and narrow; ears small; fur on nape of neck, from the back of the head to the withers, directed wholly or partially forwards; hind claws long and strong, not concealed by hair; tail evenly haired, and nearly naked at tip. Incisor teeth small, the innermost in the upper jaw shorter, rounder, and less prominent than in *Macropus*, the other two nearly equal in size, with a distinct notch near the centre of the outer edge. The fourth premolar in both jaws greatly elongated from back to front, with a ledge on its inner side, and vertical grooves on both sides. The two series of cheek-teeth nearly or quite parallel.

The Dorca Kangaroos, of which there are three species, are confined to New Guinea, and form in some respects a connecting-link between *Macropus* and the Tree-Kangaroos. With the exception that they are not arboreal, little or nothing appears to be known of the habits of these Kangaroos in a wild state.

I. BROWN DORCA. DORCOPSIS MUELLERI.

Macropus bruni (nec Schreber), Waterhouse, Jardine's Nat. Library, Mamm., vol. xi., p. 225 (1841).
Dorcopsis bruni, Schlegel and Müller, Verhandl. Nat. Ges. Nederl. Ind., p. 131 (1839-44).
Macropus mülleri, Schlegel, Ned. Tijdschr. Dierk., vol. iii., p. 353 (1866).
Dorcopsis mülleri, Garrod, Proc. Zool. Soc., 1875, p. 49; Thomas, Cat. Marsup. Brit. Mus., p. 88 (1888).

Characters.—Nose wholly naked; fur short, close, and glossy, directed forwards on nape and shoulders. General colour uniform chocolate-brown; under-parts and chin white; a faint pale cheek-stripe; ears very small and short, darker externally than crown of head; a short transverse white stripe in front of hip; inner side of hip naked; fore limbs and front of hind legs white or pale brown; back of hind legs and feet brown. Tail about equal in length to the body, brown, with the terminal three inches white, nearly naked. Length of head and body of male about 33 inches; of tail 18 inches. Female considerably smaller.

Distribution. — North-western New Guinea and adjacent islands.

The remarkable superficial resemblance between this species and *Macropus brunii*, and the confusion thereby engendered in nomenclature, has been already mentioned (p. 37).

II. GREY DORCA. DORCOPSIS LUCTUOSA.

Halmaturus luctuosus, D'Albertis, Proc. Zool. Soc., 1874, p. 110.
Dorcopsis luctuosa, Garrod, Proc. Zool. Soc., 1875, p. 49; Thomas, Cat. Marsup. Brit. Mus., p. 89 (1888).

Dorcopsis chalmersi, et *D. beccarii*, Mikl. Macl., Proc. Linn. Soc. N. South Wales, vol. ix., p. 569 (1884), vol. x., p. 146 (1885).

Characters.—Smaller than the last, with the soft and thick fur longer; hair of neck reversed. General colour dark smoky-grey, of variable shade; chin brown; under-parts grey or greyish-white. Face dark grey, with a faint pale cheek-stripe; ears larger than in *D. muelleri*, with their back thinly haired and not darker than the crown. No hip-stripe; limbs dark smoky-grey; tail shorter than body, dark brown, with the tip white and naked.

Distribution.—Eastern and South-eastern New Guinea.

Curiously enough, this species presents the same superficial resemblance to *Macropus browni* as is shown by *D. muelleri* to *M. brunii*. Whether these resemblances are examples of mimicry, and if so on which side the mimicry exists, and what may be its object, is at present quite unknown.

III. MACLEAY'S DORCA. DORCOPSIS MACLEAYI.

Dorcopsis macleayi, Mikl. Macl., Proc. Linn. Soc. N. South Wales, vol. x., p. 149 (1885); Thomas, Cat. Marsup. Brit. Mus., p. 92 (1888).

Characters.—The smallest form. Fur radiating from two centres on back of neck. General colour brownish-grey, becoming somewhat lighter on the under-parts; tip of tail white and nearly naked.

Distribution.—Southern New Guinea.

As the single specimen on which this species was established presents many features connecting the typical representative of the genus with *Macropus*, it has been suggested that it may turn out to be a hybrid between *M. browni* and *D. luctuosa*,

THE TREE-KANGAROOS. GENUS DENDROLAGUS.

Dendrolagus, Schlegel and Müller, Verhandl. Nat. Ges. Nederl. Ind., p. 138 (1839-44).

General build of ordinary proportions in regard to the relative length of the limbs, and unlike that of the Kangaroos. Naked portion of nose broad and partially covered with scattered hairs; fur on nape, and sometimes on back, directed forwards. Fore limbs strong, stout, and nearly as long as the hinder pair; hind feet broad, with the two inner toes not greatly smaller than the other two; claws stout and strong, those of the two outer toes of the hind foot as much curved as those of the fore foot. Tail very long, thickly and evenly haired. The Kangaroos being such essentially terrestrial animals, it is somewhat surprising to find certain members of the family adapted for an arboreal life. Nevertheless, four species, of which three are from New Guinea, while the fourth is an inhabitant of North Queensland, are dwellers in trees, which they climb with facility. Whether the relatively large size of the fore limbs is a feature which has been re-acquired, or whether it is a primitive feature, is a question that does not very readily admit of an answer; but the close correspondence in the structure of the feet to those of the more typical members of the family would rather seem to be in favour of regarding the relative equality of the limbs as an acquired feature.

1. QUEENSLAND TREE-KANGAROO. DENDROLAGUS LUMHOLTZI.

Dendrolagus lumholtzi, Collett, Proc. Zool. Soc., 1884, p. 387; Thomas, Cat. Marsup. Brit. Mus., p. 96 (1888).

(*Plate VII.*)

Characters.—Form stout; fur long and rather coarse, reversed from withers to crown of head. General colour pale, finely

PLATE VI

grizzled grey; chin black; chest white; flanks and rest of under-parts pale yellowish-white; face black, with a paler band across the forehead; ears with short and coarse hairs, black externally, yellow internally. Fore legs to wrist, and hind legs to ankle, pale yellow; wrists and ankles darker; toes black; tail mingled black and pale yellow, the upper surface paler than the lower, but with a darker patch near the root. Length of head and body about 26 inches; of tail about the same.

Distribution.—Northern Central Queensland, in the neighbourhood of the Herbert river.

Habits.—The following account of the habits of this species is taken from that given by Dr. Carl Lumholtz, who derived most of his information from native sources; it will probably serve for those of the genus generally. The "Bungary," as this Kangaroo is termed by the natives, although apparently far from uncommon in the scrub-clad mountainous districts, is extremely difficult to find, on account of its inhabiting the most inaccessible regions, when it can only be tracked by the aid of the natives, who must be accompanied by a well-trained Dingo. When one of these animals is put on the trail of a Bungary in the early morning, when the scent is fresh and strong, it follows up the tracks until it reaches the tree which the creature has ascended. One of the hunters then climbs the tree, and either seizes hold of the long tail of the Kangaroo with one hand, at the same battering in its head with a club held in the other, or compels the animal to leap down, when it is at once seized by the dog. It is said that two or three of these Kangaroos are frequently found asleep in a single tree; and that they are chiefly nocturnal in their habits, being especially active on moonlight nights. From the absence of claw-marks of the Bungary on the stems of all but one particular kind of tree, the creature apparently frequents only that sort, up which it may often be heard climbing at night. During the rainy season, small young trees are those especially selected. The natives report that the Bungary will jump from great heights on to the ground, when its movements are agile and rapid. From being frequently found at great distances from water, it is popularly believed not to drink. The flesh is much esteemed by the natives, and is said to be very palatable to Europeans, although frequently uninviting, owing to the presence of a parasite, which burrows beneath the skin.

II. BLACK TREE-KANGAROO. DENDROLAGUS URSINUS.

Dendrolagus ursinus, Schlegel and Müller, Verhandl. Nat. Ges. Ind., p. 141 (1839-44); Thomas, Cat. Marsup. Brit. Mus., p. 94 (1888).

Characters.—Size about the same as the last, but tail considerably longer than the head and body; form thick and clumsy; nose tuberculated inferiorly, and so thinly haired as to be practically naked; the tuberculated portion nearly white, and the hairy part black; lower lip with two grooves; fur long, straight, and coarse, except on the face, which is covered only with pale under-fur; the latter short, woolly, and pale brown; fur of back of neck directed forwards. General colour uniform black; face pale whitish-brown or grey; under-parts a pale brown; ears short, rounded, with soft woolly black hairs externally, and on the inside of the edges; tip of tail sometimes yellow.

Distribution.—North-western New Guinea.

III. BROWN TREE-KANGAROO. DENDROLAGUS INUSTUS.

Dendrolagus inustus, Schlegel and Müller, Verhandl. Nat. Ges. Ind., p. 143 (1839-44); Thomas, Cat. Marsup. Brit. Mus., p. 95 (1888).

Characters.—Larger and rather more slenderly built than the last, with the tail considerably longer than the head and body.

Agreeing with *D. lumholtzi* in having the general colour of the body, limbs, and tail grey, and the face dark, it may be distinguished therefrom by having the whole of the back and sides dark grizzled grey, instead of the back pale grey and the flanks white.

Distribution.—New Guinea.

IV. DORIA'S TREE-KANGAROO. DENDROLAGUS DORIANUS.

Dendrolagus dorianus, Ramsay, Proc. Linn. Soc. N. South Wales, vol. viii. p. 17 (1883); Thomas, Cat. Marsup. Brit. Mus., p. 98 (1888).

Characters.—A large, imperfectly-known species, with the tail considerably shorter than the head and body, characterised by the forward direction of the fur of the whole of the back. The general colour is uniform dark brown, with the face paler, and the tail black.

Distribution.—South-eastern New Guinea.

THE BANDED WALLABIES. GENUS LAGOSTROPHUS.

Lagostrophus, Thomas, Proc. Zool. Soc., 1886, p. 544.

Form as in *Macropus;* nose naked; hind feet covered with long bristly hairs, concealing the claws; back cross-banded.

BANDED WALLABY. LAGOSTROPHUS FASCIATUS.

Kangurus fasciatus, Péron and Lesueur, Voyage Terres Austr., vol. i., p. 114 (1807).
Halmaturus fasciatus, Goldfuss, Isis, 1819, p. 268.
Macropus fasciatus, Fischer, Synops. Mamm., p. 284 (1829).
Lagorchestes fasciatus, Gould, Mamm. Austral., pl. lvi. (1849).
Lagostrophus fasciatus, Thomas, Proc. Zool. Soc. 1886, p. 544; Cat. Marsup. Brit. Mus., p. 104 (1888).

Characters.—Size small; general form light and graceful; fur thick and soft. General colour grizzled greyish-brown, arranged on the hinder part of the body in black and white transverse bands,—a character by which this species may be at once distinguished from all other members of the family. Length of head and body about 18 inches; of tail 13 inches

Distribution.—West Australia.

Habits.—The following summary of the habits of this pretty little Wallaby is given by Waterhouse in the first edition of the "Naturalist's Library":—

"The Banded Kangaroo," he writes, "is found at Dirk Hartog's Island, and on one or two neighbouring islands in Shark's Bay, on the west coast of Australia. It is said to inhabit the impenetrable low thickets, formed of a species of mimosa, which are found in those islands; from these bushes it cuts away the lower branches and spines, so as to form galleries communicating one with another, and where they take refuge in time of danger."

Although abundant in the islands of Shark's Bay, Péron states that none were to be found on the mainland. These little Kangaroos, like all those feeble animals which have neither the power of attack nor of defence, are, like the Hares, extremely timid. The slightest noise causes them to take flight to the thick brushwood in which their galleries are constructed, and where it is impossible to pursue them; hence, although very common, they are difficult to procure.

The flesh of these animals is said to resemble that of the Rabbit, but has a slight aromatic flavour, arising from the nature of the plants on which they feed, nearly all of which are fragrant.

At the time that Péron visited the islands, all the females carried young in their pouch, and the courage with which they sought to save their offspring was truly admirable. Although wounded, they fled with the young in their pouch, and never left them, until, overcome with fatigue and loss of blood, they could no longer carry them; then they stopped, and, squatting themselves on the hind legs, helped the young to get out of the pouch by means of the fore feet, and sought to place them in a situation favourable for retreat.

THE RAT-KANGAROOS. GENUS POTOROUS.
Potorous, Desmarest, Nouv. Dict. d'Hist. Nat., vol. xxiv., Tabl. Méthod., p. 20 (1804).

The Rat-Kangaroos, often incorrectly spoken of as Kangaroo-Rats, are represented by several genera, and constitute the second Sub-family (*Potoroinæ*) of the *Macropodidæ*, which is distinguished from the *Macropodinæ* by the following characters :—

Size small; claws of fore feet very long, those of the three median toes disproportionately larger than those of the other toe; hind foot with only four toes; tail long and hairy. Canine teeth always present, and generally well-developed; inner upper incisor taller than either of the other two; the fourth premolar set in the same general line as the molars, or slightly bent outwards in front, and of great relative length from front to back; molars more or less tuberculate, decreasing in size from front to back of the series. The group is confined to Australia and Tasmania.

The following distinctive features apply to the genus under consideration, dividing it from the other members of this group:—

Size variable; nose naked; ears very short, rounded; fore claws long and rather slender; hind limbs not disproportionately longer than the front pair; hind feet very short, less than the length of the head, with the soles naked and coarsely granulated; tail tapering, hairy, without crest. Fourth premolar tooth of considerable length, with from two to four shallow vertical grooves.

The Rat-Kangaroos of this genus, as indicated by the small size of their hind feet, are less saltatorial in their habits than the members of the other genera of the Sub-family, and there fore depart more widely from the Wallabies. According to Gould, although these Rat-Kangaroos stand as much on their hind legs as do the other members of the Sub-family, they run in a totally different way, using both fore and hind legs in a kind of gallop, and never attempting to kick with the hind feet. Externally the three species of the genus are very similar, while they present a considerable degree of variation in the structure of their skulls.

I. BROAD-FACED RAT-KANGAROO. POTOROUS PLATYOPS.

Hypsiprymnus platyops, Gould, Proc. Zool. Soc., 1844, p. 103.
Potorous platyops, Thomas, Cat. Marsup. Brit. Mus., p. 121 (1888).

Characters.—Size very small; naked portion of nose not extending backwards along the top of the muzzle; face very short and broad; fur long, coarse, and straight; hind feet very short, long-haired. General colour dark grizzled greyish-brown; under-parts white or greyish-white; back of ears dark brown; tail black above, dirty white beneath. Length of head and body about 14 inches; of tail 8 inches.

Distribution.—West Australia.

II. GILBERT'S RAT-KANGAROO. POTOROUS GILBERTI.

Hypsiprymnus gilberti, Gould, Proc. Zool. Soc., 1841, p. 14.
Hypsiprymnus micropus, Waterhouse, in Jardine's Naturalist's Library Mamm. vol. xi., p. 180 (1841).
Potorous gilberti, Thomas, Cat. Marsup. Brit. Mus., p. 120 (1888).

Characters.—Size medium; naked portion of nose extending backwards along the top of the muzzle; face long and narrow; hind feet long, short-haired. Fur and general coloration as in last species; tail grey at base, deepening to black at tip. Length of head and body about 15½ inches; of tail 7 inches.

Distribution.—South-western Australia.

III. COMMON RAT-KANGAROO. POTOROUS TRIDACTYLUS.

Didelphis tridactyla, Kerr, Linn. Anim. Kingdom, p. 198 (1792.)
Didelphis murina, Cuvier, Table Element., p. 126 (1798).
Hypsiprymnus murinus, Goldfuss, Handb. Zool., p. 447 (1820).
Hypsiprymnus apicalis, Gould, Mamm. Austral., pl. lxviii. (1851).
Potorous rufus, Higgins and Petterd, Proc. Roy. Soc., Tasmania, 1883, p. 181.
Potorous tridactylus, Thomas, Cat. Marsup. Brit. Mus., p. 117 (1888).

(*Plate VIII.*)

Characters.—Size variable, large or medium; other external characters as in *P. gilberti*, but the naked portion of the nose extending rather further upwards on the muzzle. Length of head and body reaching to 16½ inches; of tail 9 inches.

Distribution.—New South Wales, Victoria, South Australia, and Tasmania.

Varieties.—One Tasmanian race (*P. apicalis*) attains larger dimensions than specimens from the mainland, and also has the fourth premolar tooth larger and with four instead of three grooves. There is also a dwarf Tasmanian form (*P. rufus*) inferior in size to mainland examples; but intermediate varieties appear to connect all three together.

Habits.—Like the other members of the genus, this species appears to prefer ground covered with low bush or long grass. Mainly nocturnal, it is said to be always scratching the ground in search of the roots on which it feeds, and in cultivated districts it inflicts much harm on the potato crops.

THE PLAIN RAT-KANGAROOS. GENUS CALOPRYMNUS.

Caloprymnus, Thomas, Cat. Marsup. Brit. Mus., p. 114 (1888).

Nose naked; ears short and rounded; fore claws long, and strong; hind feet longer than head, with the naked soles coarsely granulated; tail thin, cylindrical, evenly short-haired, without crest. In respect of the conformation of the skull, the single representative of this genus agrees with *Bettongia*, but the molar teeth are like those of *Æpyprymnus*, while the last premolar resembles that of *Potorous*, the nasal region of the skull being peculiar. The species is thus intermediate between the other members of the sub-family, and affords some grounds for the inclusion of the whole of them in the single genus *Potorous*.

1. PLAIN RAT-KANGAROO. CALOPRYMNUS CAMPESTRIS.

Bettongia campestris, Gould, Proc. Zool. Soc., 1843, p. 81.
Hypsiprymnus campestris, Schinz, Syn. Mamm., vol. ii., Suppl., p. 47 (1845).
Caloprymnus campestris, Thomas, Cat. Marsup. Brit. Mus., p. 115 (1888).

Characters.—Size large; form delicate and slender; face broad between the eyes; fur soft and straight. General colour grizzled grey, darker on the back, brighter on the flanks; underparts pale sandy-white; ears very close, short, with yellow hairs. Legs bright sandy rufous; feet white, short-haired;

tail thinly covered with pale yellowish hairs, closest on the under side of the tip; centre of chest naked and glandular. Length of head and body about 18 inches; of tail 14 inches.

Distribution.—South Australia.

THE PREHENSILE-TAILED RAT-KANGAROOS.
GENUS BETTONGIA.

Bettongia, Gray, Mag. Nat. Hist., vol. i., p. 584 (1837).

Nose wholly naked; ears very short and rounded; fore claws long and strong; hind feet longer than head, with the naked soles coarsely granulated; tail more or less prehensile, thickly haired, with a more or less distinctly marked crest. Last premolar tooth long and stout, with from seven to fifteen distinct oblique grooves; molars squared, with four tubercles, the last of the series much the smallest of the four.

Externally the four species of this genus are remarkably similar,—so much so that it is difficult to distinguish them by appearance alone. There are, however, very well-marked points of difference in the skulls and teeth. These creatures are the only ground-dwelling Mammals with prehensile tails; the latter being used to carry bunches of grasses and sticks, which are held by the tail being bent down over and round them.

1. LESUEUR'S RAT-KANGAROO. BETTONGIA LESUEURI.

Hypsiprymnus lesueuri, Quoy and Gaimard, Voyage Uranie, Zool., p. 64 (1824).
Hypsiprymnus grayi, Gould, Proc. Zool. Soc., 1840, p. 178.
Bettongia grayi, Gray, in Grey's Australia, Appendix, vol. ii., p. 403 (1841).
Perameles harveyi, Waterhouse, Proc. Zool. Soc., 1842, p. 47.
Bettongia lesueuri, Thomas, Cat. Marsup. Brit. Mus., p. 112 (1888).

Characters.—Fur soft, close, and thick. General colour grizzled grey; under-parts white; sometimes an indistinct white hip-mark. Legs white; feet white or pale brown, with long bristly hairs nearly concealing the claws; tail coloured above like the back, the upper hairs not forming a distinct crest, below pale brown or white, with the tip white. Length of head and body about 18 inches; of tail 12 inches.

Distribution.—South and West Australia.

Habits.—Writing more than thirty years ago, Krefft observes: "This burrowing species has long retreated before the herds of cattle with which the plains bordering on the Murray are now stocked, and it is no longer to be found south of that river,—so, at least, the natives assure me,—and whenever we went out hunting for it, we always had to cross to the New South Wales side.

"It is a truly nocturnal animal, which always leaves its burrow long after the sun is down, in fact, never before it is quite dark. I often watched near their holes, gun in hand, listening to their peculiar call; but I always had great difficulty in procuring specimens, as they were very shy, and hardly to be distinguished from the surrounding objects. The best plan is always to dig them out, an operation in which the black-fellows are very expert, though it is rather tedious work, as the holes run into each other, and, being sometimes ten feet deep, several shafts have to be sunk, before a couple of 'Boomings,' as the animals are called by the natives of the Murray district, can be secured."

Although but a single offspring is produced at a birth, the writer quoted believes that this species, and probably also its allies, breed several times during the year.

In captivity these Rat-Kangaroos are wild and intractable,

generally either escaping by burrowing, or killing themselves by running their heads against the walls of their enclosure.

II. TASMANIAN RAT-KANGAROO. BETTONGIA CUNICULUS.

Hypsiprymnus cuniculus, Ogilby, Proc. Zool. Soc., 1838, p. 63.
Bettongia cuniculus, Gould, Monogr. Macropodidæ, pl. xxix. (1842); Thomas, Cat. Marsup. Brit. Mus., p. 106 (1888).

Characters.—Fur and general colour as in last species, but no trace of a hip-mark; tail above usually as in *B. lesueuri*, but sometimes dark-brown or black towards the tip, beneath dirty white, with the tip sometimes white all round. Length of head and body about 18 inches; of tail 15 inches.

Distribution.—Tasmania.

Habits.—Feeding, like the other members of the genus, on roots and grass, this species frequents the open, sandy, or stony, forest-land of Tasmania, avoiding the thick and damp scrub.

III. GAIMARD'S RAT-KANGAROO. BETTONGIA GAIMARDI.

Kangurus gaimardi, Desmarest, Mamm., Suppl., vol. ii., p. 542 (1822).
Hypsiprymnus whitei, Quoy and Gaimard, Voyage Uranie, Zool., p. 62 (1824).
Hypsiprymnus formosus, et *H. phillippii*, Ogilby, Proc. Zool. Soc., 1838, p. 62.
Hypsiprymnus (Bettongia) gaimardi, Waterhouse, Nat. Hist., Mamm., vol. i., p. 207 (1846).
Bettongia gaimardi, Flower, Cat. Osteol., Mus. Roy. Coll. Surg. pt. ii., p. 726 (1884); Thomas, Cat. Marsup. Brit. Mus., p. 108 (1888).

Characters.—Fur more woolly than in the other species.

General coloration as in *B. lesueuri*, but the prevailing tone more yellowish or fawn; hind feet white. Terminal two-thirds of tail gradually darkening, and the hair increasing in length, till a distinct black crest is formed on the terminal third; inferior surface of tail short-haired, white. Length of head and body about 16 inches; of tail about 11 inches.

Distribution.—New South Wales.

IV. BRUSH-TAILED RAT-KANGAROO. BETTONGIA PENICILLATA.

Bettongia penicillata, Gray, Mag. Nat. Hist., vol. i., p. 584 (1837); Thomas, Cat. Marsup. Brit. Mus., p. 110 (1888).
Hypsiprymnus penicillatus, Waterhouse, in Jardine's Naturalist's Library, Mamm., vol. xi., p. 185 (1841).
Hypsiprymnus ogilbyi, Waterhouse, *loc. cit.*
Bettongia ogilbyi, Gray, in Grey's Australia, Appendix, vol. ii., p. 403 (1841).
Bettongia gouldi, Gray, List Mamm. Brit. Mus., p. 94 (1843).

(*Plate IX.*)

Characters.—Fur and general coloration not markedly different from those of the other species. Feet pale brown; bristly hairs of hind feet not concealing the claws; tail long, with a prominent black crest along the terminal third or two-thirds of the upper surface, beneath pale brown. Length of head and body about 14 inches; of tail 12 inches.

Distribution.—The whole of Australia, except the extreme north.

Habits.—These Rat-Kangaroos are social and nocturnal in their habits, and feed chiefly upon roots and grasses. The chief use of the prehensile tail seems to be for carrying bunches of grass for the construction of the nest. The latter is built in a hollow in the ground specially excavated by the animal, and has its aperture placed on the level of the herbage by which it is sur-

BRUSH-TAILED RAT-KANGAROO.

rounded, so that its detection is a matter of some difficulty. When inhabited, the entrance of the nest is closed with grass, dragged in by its occupants.

Although generally nocturnal, when disturbed, this species may occasionally be seen abroad during the day. Krefft writes: "It is not very quick, and is easily caught, even by common dogs. I have from time to time kept numbers of these animals in captivity in an enclosure of pine-logs about seven feet high, which they used to climb with a nimbleness truly astonishing, and thus often escaped. During the daytime I always noticed these creatures crouching in some corner fast asleep, with the tail brought forward between the hind legs, the head between the paws."

THE RUFOUS RAT-KANGAROOS. GENUS ÆPYPRYMNUS.

Æpyprymnus, Garrod, Proc. Zool. Soc., 1875, p. 59.

Nose partially hairy; ears rather long; front claws very long and strong; hind feet longer than the head, with the naked soles narrow and coarsely granulated; tail evenly haired, without trace of crest. Last premolar long and narrow, with some seven or eight vertical grooves, but lacking the internal basal ledge found in *Potorous*. Molar teeth oblong, less distinctly tuberculated and more ridged than in the other genera of the sub-family, and their decrease in size from the first to the last less strongly marked.

1. RUFOUS RAT-KANGAROO. ÆPYPRYMNUS RUFESCENS.

Bettongia rufescens, Gray, Mag. Nat. Hist., vol. i., p. 584 (1837).
Hypsiprymnus melanotis, Ogilby, Proc. Zool. Soc., 1838, p. 62.

Hypsiprymnus rufescens, Waterhouse, Jardine's Naturalist's Library, Mamm., vol. xi., p. 188 (1841).
Æpyprymnus rufescens, Garrod, Proc. Zool. Soc., 1875, p. 59; Thomas, Cat. Marsup. Brit. Mus., p. 103 (1888).

Characters.—The largest of the Rat-Kangaroos, easily distinguished from all the other members of the sub-family by its ruddy colour, black-backed ears, whitish hip-stripe, and hairy nose. Nose hairy for about half way down the septum between the nostrils; fur long and coarse. General colour rufescent grey; under-parts dirty white; an indistinct white stripe across the side in front of the hip; back of ears black or dark brown; outside of hind legs grey; rest of legs white; hair on back of fore paws black and coarse, partially covering the claws; feet brown or greyish-brown; central hind claw long and strong; tail thickly haired, without crest, pale grey above, white beneath. Length of head and body about 21 inches; of tail 15 inches.

Distribution.—New South Wales; fossil in the caves of the Wellington Valley, N.S.W.

Habits.—This species is extremely common in New South Wales, feeding on roots and grasses, and seeking shelter, when pursued, in hollow logs or holes in the ground, after running with great speed for a short distance, and taking a series of bounds, immense for so small an animal, during its flight. In the daytime this Rat-Kangaroo lies securely coiled up in its nest, which is formed of dry grasses, and usually placed beneath some fallen log, or under the shelter of a low bush of shrub. Occasionally the creature prefers a "form," like that of a Hare, among low herbage, but it is never by any chance found resting in the open plains, thus differing markedly from the Tasmanian Rat-Kangaroo.

In captivity this species is readily tamed, and appears to become really attached to its owner.

THE MUSK-KANGAROOS. GENUS HYPSIPRYMNODON.

Hypsiprymnodon, Ramsay, Proc. Linn. Soc., N. South Wales, vol. i., p. 33 (1876).

The third and last sub-family (*Hypsiprymnodontinæ*) of the *Macropodidæ* is represented solely by the remarkable creature known, from its strong scent, as the Musk-Kangaroo, which forms a connecting link between the present and following families. Indeed, it is somewhat difficult to decide to which of the two the species ought to be assigned, although the presence of a large pocket-like pit on the outer side of the hinder part of the lower jaw seems to be indicative of closer affinity with the *Macropodidæ* than with the *Phalangeridæ*.

The following are the distinctive features by which the sub-family differs from the two other equivalent groups:—

Claws small, weak, and of nearly equal size; first toe of hind foot well-developed and opposable to the others; tail naked and scaly. Last premolar set obliquely to the line of the molars.

The undermentioned characters, on the other hand, may be regarded rather as distinctive of the one genus *Hypsiprymnodon*.

Size very small; form Rat-like; nose entirely naked; ears large, thin, and devoid of hair; limbs of sub-equal length, and adapted for walking; first toe of hind foot long and clawless; fourth toe of same not disproportionately larger than the other; the fifth and united second and third being alike well developed; tail cylindrical and tapering, haired only at the root.

1. MUSK-KANGAROO. HYPSIPRYMNODON MOSCHATUS.

Hypsiprymnodon moschatus, Ramsay, Proc. Linn. Soc., N. South Wales, vol. i., p. 34 (1876); Thomas, Cat. Marsup. Brit. Mus., p. 123 (1888).

Pleopus nudicaudatus, Owen, Ann. Mag. Nat. Hist., ser. 4, vol. xx., p. 542 (1877).

Characters.—Fur close, crisp, and velvety; the naked and rounded ears of a blackish flesh-colour, with some hair at the base of the back. General colour finely grizzled rusty orange-grey, the orange tinge being most marked on the back. Fore feet with the toes naked and scaly, and the sole furnished with five large transversely ridged pads; in the hind feet the upper surface of fourth toe alone hairy; and the five pads on the sole transversely striated. Tail black above; paler beneath. Length of head and body about 10 inches; of tail $6\frac{1}{2}$ inches.

Distribution.—North Queensland, in the Herbert river district.

Habits.—Having very much the general appearance of a large Rat, both as regards size and form, the Musk-Kangaroo appears to be partly terrestrial and partly arboreal in its habits, while its food includes insects, roots, and fruits. According to its original describer, it frequents the densest and dampest portions of the scrubs fringing the rivers and clothing the coast-range. Although in such situations this interesting little creature is far from rare, yet its shy and retiring disposition, coupled with the density of the forest which it frequents, render it difficult to obtain. Mainly diurnal in its habits, and, when undisturbed, by no means ungraceful in its movements, the Musk-Kangaroo procures its food on the ground by turning over leaves, twigs, and stones for the sake of the insects and worms that lie concealed beneath, while roots are dug up with its claws. At times it may be seen sitting up on its haunches and munching palm-berries, which are held in the fore paws, after the manner of a Squirrel. Except in the case of parents and their young, it is rare that more than two individuals are seen in company. The breeding season takes place during the period of the rains, and the number of young produced at a birth appears to be either two or one. The strong musky odour from which the creature takes its name is perceptible in both sexes, although much more strongly developed in the female than in her consort.

THE PHALANGERS, &c. FAMILY PHALANGERIDÆ.

The second great family of the herbivorous Diprotodont Marsupials is typically represented by the creatures properly known as Phalangers, which the colonists of Australia persist in misnaming Opossums. It includes, however, several other forms, such as the Flying Phalangers and the Koala, together with the remarkable and aberrant Long-snouted

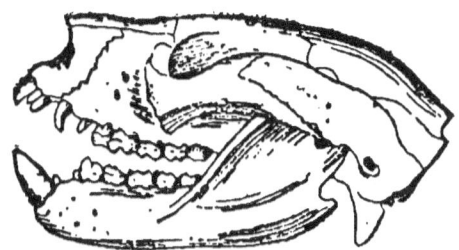

Side view of Skull of a species of Phalanger.

Pouched-Mouse. Although the number of sub-families (3), and likewise of genera (12), is the same in this family as in the *Macropodidæ*, the list of species is considerably less, being just over thirty instead of somewhat exceeding fifty.

If it were not for the existence of the Musk-Kangaroo, which, as we have seen, forms in many respects a connecting link between them, the *Phalangeridæ* would be easy enough to separate from the *Macropodidæ;* but, as it is, a clear distinctive definition of the two is by no means easy. The following are, however, the distinctive characters which serve to differentiate the members of the present family as a whole from those of the *Macropodidæ* :—

All the feet with five toes; those of the fore limbs generally sub-equal; those of the hind limbs, with the second and third united in a common integument (syndactylous), the fourth the longest of the series; the fifth but little smaller, and the first (or hallux) large, furnished with a broad clawless terminal pad, and widely opposable to the others (see fig. 3, p. 12). Tail (except in *Phascolarctus*, where it is rudimental), long, and nearly always prehensile. Stomach simple; intestine, except in *Tarsipes*, furnished with a blind appendage or cæcum; and the pouch well-developed, with its aperture directed backwards. In the skull the lower jaw lacks the large pocket-like cavity on the outer side of the hinder part, and communicating with the canal of the dental nerve, which forms so characteristic a feature in the *Macropodidæ*. Owing to the frequent presence of a variable number of functionless minute teeth in the anterior part of the jaws, the dentition as a whole is too variable to admit of precise formulation. There are, however, generally three pairs of upper and one pair of functional lower incisors; the latter wanting the scissor-like action characterising the *Macropodidæ*. Of the two or three premolars usually present, the last is generally furnished with a sharp cutting edge (although this character is less marked than in the preceding family), and is placed obliquely to the line of the molars. The molars, generally four in number, may be crowned either with blunt tubercles or with sharp-cutting crests.

It will be obvious that almost the only characters which separate the *Phalangeridæ* from the Musk-Kangaroo are the structure of the lower jaw and the backward direction of the aperture of the pouch.

As regards their distribution in space, the *Phalangeridæ* range not only over Australia and Tasmania, but are likewise found in New Guinea and the Austro-Malayan islands, their extreme westerly limit being Celebes. The two most peculiar and aberrant genera (*Phascolarctus* and *Tarsipes*) are, however, exclusively Australian.

As is well remarked by the author of the admirable technical work the "British Museum Catalogue of Marsupials," the Phalangers and their allies may be regarded as the most generalised, and therefore presumably the most ancient types of Diprotodont Marsupials at present existing. Related closely to one another in respect of external form and appearance, they differ very widely in regard to their dental characters, thus showing that specialisation has played a large part in the latter. It is scarcely necessary to observe that the numerous small functionless teeth occurring in so many members of the family are remnants of the fuller series of teeth characterising the second and more generalised primary sub-division of the Marsupials known as the Polyprotodonts. And it may be incidentally mentioned here, as tending to show the origin of the comparatively specialised Diprotodonts from the generalised Polyprotodonts that even in such a highly specialised group as the Kangaroos vestiges of at least five pairs of incisor teeth have recently been discovered in the fœtus.

In their modes of life the Phalangers and their allies are essentially arboreal creatures, the great majority of them being largely assisted in their climbing by their highly prehensile tail. Some, however, have "gone one better" than this, and have developed large parachute-like expansions of skin from the sides of the body, by means of which they are able to take long flying leaps from bough to bough, and thus from tree to tree. And it may be mentioned here as a somewhat remarkable circumstance that the different groups of these Flying Phalangers, like their analogues the Flying Squirrels, have developed their parachutes, independently of one another, from distinct groups of their non-volant cousins. In the case of the three genera of the flying forms, this is proved by the circumstance that while their relationship to one another is of the most distant nature, each is closely allied to a non-flying genus.

While the great majority of the members of the family are purely vegetable feeders, subsisting chiefly on leaves and fruits, a few feed either entirely or partially on insects, while others have taken to a diet of flesh.

THE KOALAS. GENUS PHASCOLARCTUS.

Phascolarctos, Blainville, Bull. Soc. Philom., 1816, p. 110.

The genus *Phascolarctus*, which is represented by a single Australian species, is also the type of the first sub-family (*Phascolarctinæ*), the distinctive characters of which are as follows :—

Tail wanting ; muzzle short and broad ; tongue not extensile ; cheeks furnished with pouches ; intestine with a blind appendage, or cæcum ; teeth large ; only one premolar in the upper jaw.

The genus itself may be defined by the following assemblage of characters :—

Size large; form very stout and clumsy; fur thick and woolly; ears large and thickly furred ; fore toes sub-equal, their relative lengths in the order 4, 3, 5, 2, 1, and the first and second opposable to the rest ; claws thick, strong, and sharp ; soles of both pairs of feet granulated, without striated pads. Two teats ; eleven pairs of ribs. Upper molar teeth with low squared crowns, carrying curved crests, of which the convexity is turned outwards.

KOALA

I. THE KOALA. PHASCOLARCTUS CINEREUS.

Lipurus cinereus, Goldfuss, Isis, 1819, p. 274.
Phascolarctos fuscus, Desmarest, Mamm., vol. i., p. 276 (1820).
Phascolarctos koala, Gray, in Griffith's Anim. Kingdom, p. 205 (1827).
Phascolarctos cinereus, Fischer, Syn. Mamm., p. 285 (1829); Thomas Cat. Marsup. Brit. Mus., p. 210 (1880).

(*Plate X.*)

Characters.—General colour grey; under-parts white or yellowish-white. Nose thinly clothed with minute hairs; ears rounded, white, save the hairs on their hinder surface, which are black with white tips; rump dirty white, sometimes irregularly spotted; feet white. Length of head and body about 32 inches.

Distribution.—Eastern Australia.

Habits.—Having a superficial resemblance to a small Bear, as is especially shown by the absence of a tail, the broad furry ears, short and wide head, and stout and short limbs, the Koala is commonly spoken of by the Australian colonists as the "Native Bear;" while its sluggish habits have occasionally given rise to the title of "Native Sloth." Nevertheless, the creature is a true Marsupial, and also one of a most lazy and sluggish disposition, moving about on the stems of the gum-trees in the most deliberate manner, and feeding chiefly upon leaves and grasses. Its favourite haunts are the hollow stems of trees, from which it issues forth by night, and occasionally also by day, in search of food. In the evenings, more especially during the autumn, one of these animals may frequently be observed crawling slowly along the topmost branches of some giant Eucalyptus; while if it is a female, it is as likely as not to have its solitary offspring perched securely on its back. Always, apparently, a solitary creature, the Koala moves awkwardly enough on the ground, where, if danger threatens, it always seeks safety by endeavouring to reach the nearest tree, up which it soon climbs till out of reach of gunshot. When alarmed or wounded, the Koala utters a loud, hoarse groaning cry, which can be heard at a great distance. The flesh is considered a great delicacy by the natives, and is regarded as not unpalatable even by Europeans. Of its pursuit by the natives in the neighbourhood of Port Jackson, Colonel Patterson writes as follows: "The natives examine, with wonderful rapidity and minuteness, the branches of the loftiest gum-trees, and upon discovering a Koala, they climb the tree in which it is seen with as much ease and expedition as a European would mount a tolerably high ladder. Having reached the branches, which are sometimes forty or fifty feet from the ground, they follow the animal to the extremity of a bough, and either kill it with the tomahawk or take it alive. The Koala feeds upon the tender shoots of the blue gum-tree, being more particularly fond of this than of any other food; it rests during the day on the tops of these trees, feeding at ease or sleeping. In the night it descends and prowls about, scratching up the ground in search of some particular roots."

The Koala must be an abundant animal, since from 10,000 to 30,000 skins are annually imported into London, while in 1889 the enormous total of 300,000 was reached. The value of these skins now ranges, according to Poland, from fivepence to a shilling each; and they are mainly used in the manufacture of those articles for which a cheap and durable fur is required.

THE CUSCUSES. GENUS PHALANGER.

Phalanger, Storr, Prodrom. Method. Mamm., p. 33 (1780.)

The second and chief sub-family (*Phalangerinæ*) of the

family under consideration is typically represented by the Cuscuses, and may be characterised as follows :—

Tail well-developed, generally prehensile; muzzle short and broad; tongue not extensile; no cheek-pouches; intestine provided with a large blind appendage or cæcum; stomach simple; teeth large.

In the genus *Phalanger* itself we find the following features distinctive: Size large or medium; form stout; fur thick and woolly; ears moderate or short. Front toes sub-equal, their relative lengths in the order 4, 3, 5, 2, 1; claws long, stout, and curved; soles of feet naked and striated, with large ill-defined pads. Tail strong, with its terminal portion naked, smooth or granulated, and prehensile. Teats four.

Of the five species of Cuscuses, one is common to Northern Australia, New Guinea, and the Austro-Maylayan Islands, while the others are restricted to the two latter regions, ranging as far westwards as Celebes. Completely arboreal and mainly herbivorous in their habits, the Cuscuses are slow and sleepy animals, passing the day curled up among the foliage, and only waking into activity with the approach of night. The different species present a great amount of variation as regards size, colour, and dentition, and are therefore not always easy to determine.

I. BLACK CUSCUS. PHALANGER URSINUS.

Phalangista ursina, Temminck, Monogr. Mamm., vol. i., p. 10 (1827).
Cuscus ursinus, Lesson, Man. Mamm, p. 219 (1827).
Phalanger ursinus, Thomas, Cat. Marsup. Brit. Mus., p. 195 (1888).

Sexes similar; fur coarse and harsh; general colour dark brown or black. Nose naked, with the bare portion extending in a wedge-shaped form up the muzzle nearly to the level of the eye. Face blackish, with the hairs tipped with dirty white, and a whitish patch round the eye; ears short and rounded, thickly covered with short, coarse hairs, which are dirty white internally and on the edges, and black on the outside. Chin and under-parts greyish, with the hairs black at the base, and dirty yellowish-white at the tip. Limbs black; toes naked above; soles of feet broadly striated. Tail furry like the body for about half its length, the hair extending two or three inches further on the upper than on the lower surface. Somewhat larger than the next species, with a proportionately longer tail.

Distribution.—Celebes.

Habits.—Like the other species of the genus, this Cuscus lives mainly upon the leaves of trees, of which it consumes large quantities. Although its movements are slow, the animal is difficult to kill, owing to the denseness of its soft, thick fur, which deadens the effect of a charge of shot, and also to its extreme tenacity of life, which is said to be so great that even a fracture of the spine or a perforation of the brain will often not prove fatal for many hours. By the natives of the islands which they inhabit the flesh of all the species of Cuscus is much esteemed as food; and Mr. Wallace relates an instance of the great difficulty which he experienced in obtaining a fine specimen of one of the species, until he found out that by promising to restore the body he would be readily permitted to retain the skin. Many of these animals are taken by the natives, who climb after them among the branches, where, from their slow motions, they are easily made prisoners. All exhale a strong odour, which in some instances is stated to be so powerful as to permeate the whole woods in which they dwell.

II. SPOTTED CUSCUS. PHALANGER MACULATUS.

Phalangista maculata, Geoffr., Cat. Mus., p. 149 (1803).

Phalangista papuensis, Desmarest, Mamm., Suppl., vol. ii., p. 541 (1822).
Cuscus maculatus, Lesson and Garnot, Voyage Coquille, Zool., vol. i., p. 150 (1826).
Phalanger maculatus, Thomas, Cat. Marsup. Brit. Mus., p. 197 (1888).

Sexes usually different, the females being larger than the males. Size large; fur soft; top of muzzle above nose thinly covered with hair; ears small, thinly clothed on both sides with soft woolly hair. General colour consisting of various combinations of white, rufous, and black; under-parts white tinged with yellow or rufous; tail generally deep yellow, furry from one-half to three-quarters of its length above, and from one-third to one-half beneath. Length of head and body about 26 inches; of tail 19 inches. The yellow colour of the base of the tail will serve to distinguish the dark varieties of this species from the preceding, in which the same region is dark like the body; these two species being the only members of the genus in which the ears are thickly haired both externally and internally.

Distribution.—Northern Australia, in the Cape York district, Southern New Guinea, and the Austro-Malayan Islands, eastwards from Saleyer, but unknown in Celebes, the Southern Moluccas, or the Halmahéra group.

As a rule, the females are generally grey and black, while the males are usually spotted, although sometimes they resemble the ordinary grey female, with the exception of having a few indistinct whitish spots on the back and flanks. From this form there is a gradual transition to one in which the fur is nearly wholly white, save for a few dark spots. Other males assume a more or less decided red tinge; the rufous tint sometimes occupying only a part or the whole of the portions of the fur which are usually dark, but in other cases suffusing those regions of the body which are generally white. The most curious feature about the coloration of the species is, however, that the females inhabiting the small island of Waigiou, to the south of Ceram, are coloured like the fully spotted and reddish males of the same island. That the ordinary grey hue of the female is the primitive coloration of the species, may be pretty safely admitted; the spotted and rufous males being a higher and more specialised development of coloration. For some unknown and at present apparently inexplicable reason this specialised type of coloration has spread to the females in the island above-mentioned, which thus contains what we may call the most advanced representatives of the species.

Habits.—In Australia this Cuscus is described as a shy, solitary creature, which is but rarely seen, although often more frequently observed by day than by night. It appears to be sparingly distributed over the thin bush, especially in the neighbourhood of the creeks and swamps, where it is generally found singly. Although it is probable that the chief food of this Cuscus is, like that of its allies, of a vegetable nature, yet the creature bears an ill-repute among the colonists, by whom it is commonly termed the "Tiger Cat," on account of its alleged depredations on the poultry-roost.

III. GREY CUSCUS. PHALANGER ORIENTALIS.

Didelphis orientalis, Pallas, Misc. Zool., p. 59 (1766).
Phalanger orientalis, Storr, Prodrom. Method. Mamm., p. 33 (1780); Thomas, Cat. Marsup. Brit. Mus., p. 201 (1888).
Phalangista rufa, et *P. alba*, Geoffr., Cat. Mus., pp. 148, 149 (1803).
Cuscus orientalis, Gray, List Mamm. Brit. Mus., p. 84 (1843).

Characters.—In common with the two following members of

the genus, this species differs from the two foregoing, in having the inside of the ears nearly or quite naked.

Rather inferior in size to the last, and the females smaller than the males; fur soft and woolly, very variable in length. General colour grey, but varying from nearly white to dark greyish-brown, albino individuals, nearly always males, being comparatively common. Upper-parts uniform in tint; the head, back, and outer surface of limbs being of the same hue, which is generally decidedly paler in the males. Upper surface of muzzle naked for about half the distance to the eyes. The internally naked ears small, and round, well furred over for the greater part of their external surface. Chin and underparts generally pale grey or white; but in some (mostly male) individuals the throat and neck strongly suffused with yellow or rufous. Tail usually haired for its basal half above and its basal fourth below, but very variable in this respect.

Distribution.—Timor, Bouru, Sula, Guébeh, and the islands eastwards to New Guinea.

Variety.—In the New Britain group and the Solomon Islands, as far eastwards as San Christoval, this species is replaced by a considerably smaller race (*P. breviceps*), in which the general colour is usually darker, and the dark line down the back less distinctly marked than in the typical form.

Habits.—Writing of the habits of the Moluccan representatives of the genus, Mr. H. O. Forbes observes that these animals "are very plentiful, and in May the females all seem to have a little one in their pouch. One of them was a tiny creature, about two inches long, quite hidden in its pouch, fixed by its lips formed into a single round orifice to its mother's teat. They are much eaten by the natives, by whom they are caught in nooses set in the trees, or by artifice.

In moonlight nights, creeping stealthily to the foot of a tree, where they have observed one sleeping, taking care not to lift their heads so that the light flash in their eyes, they imitate at short intervals its cry by placing the fingers in the nose; the Cuscus descends, and is fallen on by the watchers below."

IV. WALLACE'S CUSCUS. PHALANGER ORNATUS.

Cuscus ornatus, Gray, Proc. Zool. Soc., 1860, pp. 1 and 374.
Phalanger ornatus, Thomas, Cat. Marsup. Brit. Mus., p. 205 (1888).

Characters.—This species, which was discovered by Dr. A. R. Wallace, may be distinguished from the last by its smaller size and lighter build, and by the back being more or less distinctly spotted with white, the dark line down the latter being well marked. Old males frequently become highly rufous on the fore quarters, neck, and under-parts.

Distribution.—Molucca islands, viz., Morotai, Ternate, Batchian, and Halmahéra.

V. CELEBEAN CUSCUS. PHALANGER CELEBENSIS.

Cuscus celebensis, Gray, Proc. Zool. Soc., 1858, p. 105.
Phalangista celebensis, Jentink, Notes Leyden Mus., vol. v., p. 181 (1883).
Phalanger celebensis, Thomas, Cat. Marsup. Brit. Mus., p. 206 (1888).

Characters.—The smallest of the genus. Fur thick and soft; colour uniform grey, with a coppery tinge, but no trace of a dark line down the back. Top of muzzle naked nearly to the level of the corner of the eye. Ears larger than in the other species, covered behind with soft fawn-grey hairs, but with their margins and inner surface naked. Chin and under-parts white or yellowish-white, sometimes passing into deep chestnut at and near the root of the tail. Tail furry for about one-half of its length superiorly, and for two-fifths below; its terminal portion with more fine hairs scattered over its upper and lateral surfaces than usual, and with the contrast in colour between the upper and under surfaces very strongly marked.

Distribution.—Celebes and Sanghir Islands.

THE TRUE PHALANGERS. GENUS TRICHOSURUS.

Trichosurus, Lesson, Dict. Class. d'Hist. Nat., vol. xiii., p. 332 (1828).

Size large; form stout; fur thick and woolly; ears medium or short. Fore toes sub-equal, their respective lengths in the order 4, 3, 2, 5, 1; claws large and strong; soles of hind feet thickly haired beneath the heels, elsewhere naked, with low, rounded, ill-defined pads. Tail strong, its terminal third or half naked inferiorly, and its extreme tip completely bare.

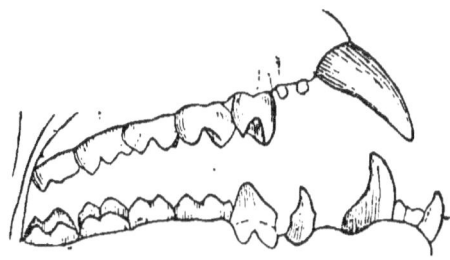

Side View of Upper and Lower Jaw of a species of *Phalanger*.

Chest with a gland. The molar teeth with four cusps which tend to form two incomplete transverse ridges; and the fourth premolar large, obliquely placed, and vertically ridged (especially in the lower jaw), thus approximating very closely to the corresponding tooth of *Hypsiprymnodon*.

Widely different in the structure of its skull and teeth from the following genus, the present one, remarks Mr. Thomas, is "not easily definable externally. Its fore feet appear, however, to be of more normal construction than in *Pseudochirus;* its tail is more densely haired terminally, although in this character it is approached by *Pseudochirus lemuroides*; and in most specimens the discoloration of fur caused by the chest-gland forms an easy method of recognising its members."

I. COMMON PHALANGER. TRICHOSURUS VULPECULA.

Didelphis vulpecula, Kerr, Linn. Anim. Kingdom, p. 198 (1792).
Didelphis vulpina, Meyer, Uebers. Zool. Entd. Neuholl., p. 23 (1793).
Phalangista vulpina, Desmarest, Nouv. Dict. d'Hist. Nat., vol. xxv., p. 475 (1817).
Trichosurus vulpecula, Jentink, Notes Leyden Mus., vol. vii., p. 21 (1884); Thomas, Cat. Marsup. Brit. Mus., p. 187 (1888).

(*Plate XI.*)

Characters.—Size small; fur close, thick, and woolly. General colour clear grizzled grey; chin more or less blackish; underparts white or dirty yellow, with a median rusty-red patch on the chest in adults. Ears long and narrow, much longer than broad, nearly naked inside, and near the tip externally. Feet white, grey, or brown. Tail thick, cylindrical, and bushy, with the terminal half or two-thirds grey, the end black, and the extreme tip occasionally white; the naked inferior portion from three to six inches in length, and transversely wrinkled. Length of head and body about 18 inches; of tail 11 inches.

Distribution.—The whole of Australia, with the exception of the Cape York district.

Variety.—The Tasmanian Phalanger (var. *T. fuliginosus*) is of larger size and stouter and heavier build than the typical form,

PLATE XI

COMMON PHALANGER.

THE TRUE PHALANGERS.

while the fur is longer and thicker; the general colour is rufous grey or deep umber-brown, and the ears have little or no white behind. The thick and bushy tail is almost entirely black. Length of head and body about 23 inches; of tail 15 inches.

Habits.—The Common Phalanger, which may be compared in size with an ordinary Domestic Cat, is an inhabitant of the tallest trees of the Australian and Tasmanian forests, preferring gum-trees to all others, and selecting for its habitation those with a hollow trunk or arms, in the recesses of which it can pass the day securely wrapt in slumber. At night the Phalanger leaves its lair to wander stealthily among the branches, upon the buds, fruits, and leaves of which it feeds. Occasionally these animals descend to the ground, perhaps in search of water; and throughout these arboreal rambles they are much assisted in climbing by their highly prehensile tail, which is alone amply sufficient to support the weight of the body. Indeed, if a dying Opossum, in falling from the bough on which it is sitting, happens to catch its tail round another branch, it will die in that position, and there hang. Phalangers are more numerous in some localities than in others, and generally frequent the neighbourhood of water. In such favoured situations one or more of these animals will certainly be found, although it requires a practised eye to detect their presence in the shades of evening or by moonlight. The eye must then scan each bough with the moon behind it, when the upright ears of the Phalanger will often betray its presence as it lies stretched along the branch, or partially concealed in a cleft. During the pairing season, and less commonly at other times of the year, the Phalanger utters a loud chattering scream, which may be heard for a considerable distance through the silent forest. Breeding but once in the year, the female Phalanger usually produces but one young at a birth, which at first is of a reddish hue; but occasionally a pair may be found in the pouch.

Since the highly aromatic leaves of the peppermint-gum form the favourite food of these animals, their flesh is naturally strong and rank; so much so, indeed, that it is almost un-eatable by Europeans, although by the natives it is regarded as the best of food.

Phalangers are much sought after by Europeans for the sake of their beautiful skins, which go to form the well-known "Opossum"-rugs. The creatures are generally obtained with the aid of a dog, which either puts them up when on the ground, till they take refuge in the nearest tree, or marks the tree in which one of them is resting. When "treed," the Phalanger generally runs only a few feet up the trunk, where it sits hissing and spitting at the dog, who remains barking furiously at the foot, until the hunters come up and despatch their quarry. When wounded, a Phalanger is a dangerous beast, and must be approached with caution. It is stated that, when freshly killed, the fur is apt to come off these animals in handfuls if they are pulled about, and many skins are irretrievably damaged in this manner.

The total number of Phalanger skins of various kinds sold in London during the year 1891 reached, according to Poland, three millions, of which the greater number belonged to the common species of the mainland. Of the ordinary variety, the value per skin varied from twopence to one shilling and five-pence, according to size, colour, quality, and the demand. The skins of the Tasmanian variety, which are now scarce, owing to restrictions on the slaughter of the animal, are, however, considerably more valuable, ranging from one shilling and fourpence to half-a-crown each. By the trade Phalanger skins are also divided into several kinds, according to the locality from whence they come. Many skins are spoiled by being worn just above the tail, owing to the animals rubbing them against the branches of the trees. To make a good "Opossum".

rug about eighty skins are required; while it is said that as many as one hundred and twenty have been used.

Writing of some examples of this species formerly kept in the Zoological Society's Gardens, Waterhouse observes that "during the daytime they were usually asleep, but towards evening they became active, and on the alert for their food, consisting of bread-and-milk and various vegetable substances, including fruits. Whatever eatable was given to them was taken by and held between the hands in the same manner as a Squirrel holds a nut. Occasionally a dead bird was given to these animals, which were evidently fond of such food, and most particularly the brain, which was the first part consumed."

II. SHORT-EARED PHALANGER. TRICHOSURUS CANINUS.

Phalangista canina, Ogilby, Proc. Zool. Soc., 1835, p. 91.
Trichosurus caninus, Thomas, Cat. Marsup. Brit. Mus., p. 191 (1888).

Characters.—Size large; fur comparatively short; ears short, evenly rounded, not so long as broad. General colour clear grizzled grey or deep umber-brown with a rufous tinge, paler on the fore quarters, flanks, and under-parts, darker on the hinder part of the back. Tail very thick and bushy, nearly completely black. Length of head and body about 22 inches; tail 15 inches.

Distribution.—New South Wales and Southern Queensland.

Although closely allied to the preceding form, and more especially to the Tasmanian variety, this Phalanger is readily distinguished by the markedly smaller size of its ears. It is said to differ from the common species by inhabiting only the scrub, and never frequents the open forest; but whether there are other differences in its habits has not been recorded.

THE RING-TAILED PHALANGERS. GENUS PSEUDOCHIRUS.

Pseudochirus, Ogilby, Proc. Zool. Soc., 1836, p. 36.

Size large or medium; fur short and rather woolly; ears medium or short, hairy behind. Fore toes sub-equal, the two inner ones markedly opposable to the other three; their relative lengths in the order 4, 3, 5, 2, 1. Claws moderate; soles of both fore and hind feet naked, with large rounded and striated pads. Tail long and tapering, markedly prehensile, and its tip naked inferiorly for a variable distance. Four teats. The upper molar teeth are large and oblong, with the rounded tubercles of the preceding genus modified into sharp cusps with curved ridges radiating from them; those of the lower jaw having a nearly similar structure.

The Ring-tailed Phalangers, of which there are half-a-score of species, while agreeing with the True Phalangers in their arboreal and leaf-eating habits, have a wider geographical range, extending over New Guinea, as well as Australia and Tasmania. According to the author of the British Museum Catalogue of Marsupials they naturally fall into three groups, of which the common Ring-tailed Phalanger, D'Albertis' Phalanger, and the Hoary Phalanger, may respectively be regarded as typical. Of these, the first group is confined to Australia and Tasmania, and the third to New Guinea, while the two representatives of the second group respectively inhabit North Australia and New Guinea. With the exception of the common species, all the members of the genus appear to have relatively small geographical ranges.

From the crescent-like structure of the crowns of the molar teeth—a feature in which they make some approach to the Koala—it may be pretty confidently considered that these animals are somewhat more specialised than the True Phalangers; although it should be observed that there is an approximation to the same structure in the corresponding teeth of young Cuscuses.

I. SOMBRE PHALANGER. PSEUDOCHIRUS LEMUROIDES.

Phalangista lemuroides, Collett, Proc. Zool. Soc., 1884, p. 385.
Pseudochirus (Hemibelideus) lemuroides, Collett, Zool. Jahrbuch, vol. ii., p. 923 (1887).
Pseudochirus lemuroides, Thomas, Cat. Marsup. Brit. Mus., p. 170 (1888).

Characters.—Size rather large ; fur soft and woolly ; general colour dark brownish-grey ; under-parts dirty yellowish-grey ; ears of moderate length. Limbs dark brown, becoming black terminally ; tail rather short, clothed with uniform thick black fur, the naked inferior portion short. Length of head and body about 15 inches ; of tail 12 inches.

This and the next four species constitute the first group of the genus, in which the ears are of medium length, not excessively short, and longer than broad ; the tail being tipped with white, except in the present species.

Distribution.—Central Queensland, in the Herbert river district.

This species, which is known to the natives by the name of "Yappi," is said to be fairly abundant in some portions of its range, although it has but recently been made known to science.

II. HERBERT-RIVER PHALANGER. PSEUDOCHIRUS HERBERTENSIS.

Phalangista herbertensis, Collett, Proc. Zool. Soc., 1884, p. 383.
Pseudochirus mongan, De Vis, Proc. Linn. Soc., New South Wales, ser. 2, vol. i., p. 1130 (1888).

Pseudochirus herbertensis, Collett, Zool. Jahrbuch, vol. ii., p. 917 (1887); Thomas, Cat. Marsup. Brit. Mus., p. 170 (1888).

Characters.—Size moderate; fur thick, close, and woolly. General colour dark umber-brown; chin brown; under-parts white, or greyish-white, with irregular white patches. Ears short. Limbs dark brown, sometimes ringed with white. Hair of tail woolly; from one to three inches of tip of tail white; naked portion of under-surface from five to six inches in length, coarsely shagreened. Length of head and body about 14 inches; of tail 13 inches.

Distribution.—Central Queensland, in the Herbert river district.

This species—the "Uta" of the aborigines—is remarkable for the inconstancy of the presence of the white rings on the limbs; the uniformly-coloured variety having been regarded as a distinct species, under the name of *P. mongan*. It is stated to inhabit only the very summits of the mountain ranges of the Herbert river district, and consequently nothing has been ascertained as to its habits, which are, however, doubtless like those of the other species.

III. COMMON RING-TAILED PHALANGER. PSEUDOCHIRUS PEREGRINUS.

Didelphis peregrinus, Boddaert, Elench. Animal.; vol. i., p. 78 (1785).
Didelphis caudivolvula, Kerr, Linn. Anim. Kingdom, p. 196 (1792).
Didelphis novæ-hollandiæ, Bechstein, Uebers. Vierf. Thiere, vol. ii., p. 348 (1800).
Phalangista convolutor, Schinz, in Cuvier's Thierreichs, vol. i., p. 258 (1821).

PLATE XII

COMMON RING-TAILED PHALANGER

Phalangista banksii, Gray, Ann. Mag. Nat. Hist., vol. i., p. 107 (1838).
Phalangista lanuginosa, Gould, Mammals of Australia, vol. i., pl. xx (1858).
Pseudochirus caudivolvulus, Jentink, Notes, Leyden Mus., vol. vii., p. 22 (1884).
Pseudochirus peregrinus, Thomas, Cat. Marsup. Brit. Mus., p. 172 (1888).

(*Plate XII.*)

Characters.—Size large; fur shorter than in the preceding species. General colour grey or rufous in variable proportions; under-parts white, greyish-white, or rufous; region round the eyes frequently conspicuously rufous; ears rather large, their backs generally grey anteriorly, with the posterior white patch distinct, sometimes uniformly rufous. Outer side of limbs rufous; feet white or pale rufous; tail with the middle third black or nearly so, from one to four inches of the tip white, naked inferior portion from one to four inches in length, smooth, and transversely striated. Length of head and body about 16 inches; of tail 14 inches.

Distribution.—Eastern Australia, from Southern Queensland to South Australia.

Habits.—This species, which rejoices in a number of synonyms, and was long confounded with *P. cooki*, is an animal scarcely more than half the size of the common True Phalanger. Nowhere so common as the latter, and but seldom met with in the gum-trees, the Ring-tailed Phalanger generally frequents the so called tea-tree scrub, where it lives in small colonies, and constructs a nest not unlike that of the common Squirrel. Although there is usually but a single offspring produced at a birth, it is stated that as many as three young may occasionally be found in the pouch of the female. The flesh is much less rank, and therefore far more palatable than that of the True Phalangers. From Adelaide it is stated that from two to three thousand skins of this species are annually exported to London.

IV. WESTERN RING-TAILED PHALANGER. PSEUDOCHIRUS OCCIDENTALIS.

Pseudochirus occidentalis, Thomas, Cat. Marsup. Brit. Mus., p. 178 (1888).

Characters.—Size moderate. General colour deep smoky grey; under-parts white. Ears thinly haired, with the white spot on the hinder margin small; feet darker than the rest of the limbs; tail with the white tip extending to a length of five or six inches, and the naked inferior portion smooth and about four inches in length. Length of head and body about 13 inches; of tail 12 inches.

Distribution.—Western Australia.

This species may be regarded as the western representative of the preceding one.

V. TASMANIAN RING-TAILED PHALANGER PSEUDOCHIRUS COOKI.

Phalangista cooki, Desmarest, Nouv. Dict. d'Hist. Nat., vol. xxv., p. 476 (1817).
Phalangista viverrina, Ogilby, Proc. Zool. Soc., 1837, p. 131.
Pseudochirus cooki, Thomas, Cat. Marsup. Brit. Mus., p. 176 (1888).

(*Plate XIII.*)

Characters.—Size moderate; fur very close, thick, and woolly. General colour dark smoky brown; under-parts white. Ears large, rounded, the anterior part of the back brown, the pos-

PLATE

TASMANIAN RING-TAILED PHALANGER

terior margin usually white; feet dark brown or black; tail dark brown, with from two to four inches of the tip white, and the naked inferior portion smooth, and from three to five inches in length. Length of head and body about 14 inches; of tail nearly the same.

Distribution.—Tasmania.

While the common Ring-tailed Phalanger, with which, as already mentioned, the present species was largely confounded, was discovered during Cook's first voyage, the one under consideration was obtained in the third expedition of the great navigator.

VI. YELLOW PHALANGER. PSEUDOCHIRUS ARCHERI.

Phalangista (Pseudochirus) archeri, Collett, Proc. Zool. Soc., 1884, p. 381.
Pseudochirus archeri, Collett, Zool. Jahrbuch, vol. ii., p. 912 (1887); Thomas, Cat. Marsup. Brit. Mus., p. 177 (1888).

This and all the remaining members of the genus differ from the five preceding species in the shortness of the ears, which are broader than long; while, with the exception of the present species, the tail is not tipped with white.

Characters.—Size moderate; fur soft, close, and thick. General colour grizzled greyish-green; chin greyish-white; under-parts pure white; a distinct pale yellow spot above, and another below the eyes. Ears very short, broader than long, rounded, their posterior edges and a spot beneath the base white. Nape and back with a dark median line, bordered by two indistinctly dark-edged whitish lines. Tail with the terminal third white, and the inferior naked portion less than half the total length. Length of head and body about 14 inches; of tail 13 inches.

Distribution.—Central Queensland, in the Herbert river district.

Habits.—This, the third species of Phalanger made known to science from the Herbert river district, and described by Dr. Collett in 1884, is termed by the aborigines "Tula." Mr. Carl Lumholtz states : "It is not uncommon in the upper part of the mountainous scrubs, where " it seems to be more commonly distributed than *P. herbertensis* and *P. lemuroides*, though it never goes far down the mountains. Besides being like the other Phalangers, a night-animal, it is in activity for a great part of the day, as I have seen myself. The blacks kill it by climbing up the tree and throwing sticks at it, which is often very troublesome work. The animal is not very shy, but when disturbed it runs away quickly from tree to tree, so that a black man will sometimes have difficulty in killing it, if he does not get two or three of his comrades to meet it in different trees."

VII. D'ALBERTIS' PHALANGER. PSEUDOCHIRUS ALBERTISI.

Phalangista (Pseudochirus) albertisi, Peters, Ann. Mus. Genov., vol. vi., p. 303 (1874).
Pseudochirus albertisi, Jentink, Notes Leyden Mus., vol. vi., p. 109 (1884); Thomas, Cat. Marsup. Brit. Mus., p. 178 (1888).

Characters.—Size rather less than in the last species, with a relatively shorter tail ; form stout and clumsy ; fur very long, soft, and thick. General colour shining coppery brown ; chin greyish ; under-parts pure white in the middle line, greyish laterally ; no light markings in the neighbourhood of the eyes or on the ears. The short and rounded ears thick and fleshy, covered on the back with short, soft, reddish fur; a distinct narrow black line down the nape and back. Tail thick, tapering, and woolly to the tip, gradually darkening from root to tip, with the sharply defined naked inferior portion extending nearly half the length, and coarsely granulated in old examples.

Distribution.—New Guinea.

VIII. SCHLEGEL'S PHALANGER. PSEUDOCHIRUS SCHLEGELI.

Pseudochirus schlegeli, Jentink, Notes Leyden Mus., vol. vi., p. 110 (1884); Thomas, Cat. Marsup. Brit. Mus., p. 180 (1888).

Characters.—Size considerably less than in any of the preceding species; fur thick and woolly; form light and slender. General colour dull silvery grey; under-parts yellow, tinged with rufous; stripe on back inconspicuous; fur coloured like back; no streak on forehead; a pale spot below the ear. Tail coloured like the back, minus the rufous tinge, with the extreme tip naked all round, and thus different from that of every other member of the genus.

Distribution.—North-western New Guinea.

IX. HOARY PHALANGER. PSEUDOCHIRUS CANESCENS.

Phalangista (Pseudochirus) canescens, Waterhouse, Nat. Hist. Mamm., vol. i., p. 305 (1846).
Phalangista bernsteini, Schlegel, Ned. Tijdschr. Dierk., vol. iii., p. 357 (1866).
Phalangista grisescens, Peters, Ann. Mus. Genov., vol. vi., p. 303 (1874).
Pseudochirus canescens, Thomas, Cat. Marsup. Brit. Mus., p. 181 (1888).

Characters.—May be distinguished from the last species as follows: Tail naked on under surface of tip for two or three inches; face paler than back, brownish-yellow; forehead with a dark median streak; no pale spot below ear.

Distribution.—North-western New Guinea.

X. FORBES'S PHALANGER. PSEUDOCHIRUS FORBESI.

Pseudochirus forbesi, Thomas, Ann. Mag. Nat. Hist., ser. 5, vol. xix., p. 146 (1887); *id*. Cat. Marsup. Brit. Mus., p. 183 (1888).

Characters.—May be distinguished from the last by the absence of a dark streak on the forehead, and the presence of a pale spot below the ear.

Distribution.—South-eastern New Guinea.

Writing of the type specimen, its describer observes:— "This very handsome species, the smallest of the genus, is the only one that has yet been discovered anywhere in New Guinea, except in the north-west. While agreeing very closely with *P. canescens* in many of its characters, and forming, with that species and *P. schlegeli*, a very well-defined section of the genus, it yet differs remarkably from all in the total suppression of its posterior incisors and anterior premolars." Other specimens are, however, urgently needed in order to determine whether these dental peculiarities are constant.

THE TAGUAN PHALANGERS. GENUS PETAUROIDES.

Petauroides, Thomas, Cat. Marsup. Brit. Mus., p. 163 (1888).

Size large; fur very long, soft, and silky; ears very large, oval, naked inside, hairy externally; flanks with a parachute-like expansion of skin. Claws very long, highly curved, and sharply pointed; tail long, cylindrical, and evenly bushy, with the extreme tip prehensile and naked inferiorly. Skull and molar teeth as in *Pseudochirus*.

Although very different in external appearance to the Ring-tailed Phalangers, the structure of the skull and teeth in the single representative of this genus clearly indicates that it must

PLATE XIV

TAGUAN FLYING PHALANGER

be regarded as a specially modified and volant relative of that group.

1. TAGUAN FLYING PHALANGER. PETAUROIDES VOLANS.

Didelphis volans, Kerr, Linn. Anim. Kingdom, p. 199 (1792).
Phalanger volans, Lacépède, Mém. Inst., vol. iii., p. 491 (1801).
Petaurus taguanoides, *P. macrurus*, et *P. peronii*, Desmarest, Nouv. Dict. d'Hist. Nat., vol. xxv., pp. 400, 402, 404 (1817).
Petaurista taguanoides, Desmarest, Mamm., vol. i., p. 269 (1820).
Petauroides volans, Thomas, Cat. Marsup. Brit. Mus., p. 164 (1888).

(*Plate XIV.*)

Characters.—Fur long, soft, and fluffy. General colour dark ashy-grey, varying from nearly black to pale whitish-grey. Ears very large, oval, and evenly rounded, with the inner surface completely naked, and the back covered with fur like that on the head. Under-parts white or pale yellowish. Limbs black or dark brown externally, white or pale grey on the inner surface; feet thickly fringed with black hairs; toes very thick; soles naked, with low, finely striated, and rounded pads. Tail ashy-grey or blackish, generally darkest terminally, with a short naked portion, which passes gradually into the haired region, and its surface less roughened. Length of head and body about 17 inches; of tail 20 inches.

Distribution.—Eastern Australia, from Queensland to Victoria.

Variety.—Replaced in Central Queensland by a variety (*P. minor*) differing from the typical form by its inferior size, and the feeble development of the upper canine and first premolar teeth, the latter being minute, or even absent. Length of head and body about 12 inches; of tail 18 inches.

Habits.—Like the other Flying Phalangers, this species possesses the power of taking, with the aid of its parachute, long flying leaps from tree to tree. The leap is always in a downward direction, and may be described as a kind of floating through the air, which must not by any means be confounded with the true flight of a Bat or a Bird, which can be sustained for an indefinite length of time, and is accompanied, and indeed produced, by rapid movements of the fore limbs. These animals are generally found in hilly districts, where gum-trees do not grow, and pass their whole time in the trees, feeding on the leaves and fruit, and spending the day in some hollow branch or within the stem itself, whence they issue forth for their nocturnal flight. When disturbed, or in flight, they utter a loud piercing scream, audible for a long distance.

THE STRIPED PHALANGERS. GENUS DACTYLOPSILA.

Dactylopsila, Gray, Proc. Zool. Soc., 1858, p. 109.

Size medium; ears oval, nearly naked terminally; no parachute-like membrane. Fore toes very unequal, the fourth much the longest, the others in the order 3, 5, 2, 1; fourth and fifth hind toes much longer than the others; a prominent pad on the wrist; claws long; tail long, cylindrical, evenly bushy, with the extreme tip naked inferiorly. Body conspicuously striped with black and white. Molar teeth oblong, with four cusps.

The two species of this genus, which ranges from Northern Australia to New Guinea and the Aru Islands, are readily distinguished from all their allies by the great elongation of the fourth toe of the fore limb. It has been suggested that this toe is elongated for the purpose of extracting insects and their larvæ from holes in decayed wood and from beneath the bark of trees, and consequently that the creatures are mainly or entirely insectivorous. Others consider, however, that they are leaf-eaters, like the rest of their tribe; but as a matter of fact, beyond the circumstance that they are arboreal, nothing definite appears to have been ascertained as to the habits of these peculiar Phalangers.

I. STRIPED PHALANGER. DACTYLOPSILA TRIVIRGATA.

Dactylopsila trivirgata, Gray, Proc. Zool. Soc., 1858, p. 111; Thomas, Cat. Marsup. Brit. Mus., p. 160 (1888).

Dactylopsila albertisi, Peters and Doria, Ann. Mus. Genov., vol. vii., p. 542 (1875).

Phalangista (*Dactylopsila*) *angustivittis*, Peters and Doria, *op. cit.*, vol. xvi., p. 674 (1881).

Characters.—Fur close, thick, and woolly, but rather harsh. General colour white with black stripes; chin with a black spot; under-parts and inner sides of limbs white or pale yellow. The upper surface with three black stripes, the middle one running from the back of the head along the tail, widest centrally; the two lateral ones commencing on the sides of the muzzle, passing along the neck and back, and sending branches downwards in front of the shoulders, and along the limbs to the brown feet. Soles of feet finely granulated; the pads, save the one on the wrist, which is narrow and smooth, large, rounded, and finely striated. Terminal third of tail either wholly black or with a white tip, its naked portion more than an inch in length. Length of head and body about 12 inches; of tail 13 inches.

Distribution.—Central Queensland to New Guinea, and Aru Islands.

II. MILNE-EDWARDS' STRIPED PHALANGER. DACTYLOPSILA PALPATOR.

Dactylopsila palpator, Milne-Edwards, Centenaire Mém. Soc. Philom. 1888, p. 173*.

Characters.—Distinguished from the last by the still greater elongation of the fourth toe of the fore paws.

Distribution.—South New Guinea.

THE FLYING PHALANGERS. GENUS PETAURUS.

Petaurus, Shaw, Nat. Miscell., vol. ii., pl. lx. (1791).

Size medium or small; fur very soft and silky; ears fairly large, oval, nearly naked; flanks with a broad parachute-like expansion of skin; front toes gradually increasing in length from the first, the fifth in the larger species being the longest, and the fourth in the smaller ones; claws very strong, sharp, and highly curved; tail long, evenly bushy throughout. Glands present on chest and crown of head. Molars square, with rounded corners, and furnished with four cusps, except the last, which is triangular.

The three species of this genus range over New Guinea and part of Australia, including the area from the Halmahéra group of islands to Victoria. As mentioned under the head of the latter, these Flying Phalangers appear to be descended from the next genus, or an allied extinct form.

1. SQUIRREL FLYING PHALANGER. PETAURUS SCIUREUS.

Didelphys sciurea, Shaw, Zool. New Holland, vol. i., p. 29 (1794).

Petaurus sciureus, Desmarest, Nouv. Dict. d'Hist. Nat., vol. xxv., p. 403 (1817); Thomas, Cat. Marsup. Brit. Mus., p. 153 (1888).

Petaurista sciurea, Desmarest, Mamm., vol. i., p. 270 (1820).

Phalangista sciurea, Schinz, Cuvier's Thierreichs, vol. i., p. 260 (1821).

Belideus sciureus, Lesson, Nouv. Tabl. Règne Anim., Mamm., p. 189 (1842).

PLATE XV.

SQUIRREL FLYING-PHALANGER

Belideus gracilis, De Vis, Proc. Linn. Soc. N. South Wales, vol. vii., p. 619 (1882).

(*Plate XV.*)

Characters.—Size medium ; fur soft and silky, slightly woolly. General colour soft pale grey, with a well-defined dark brown or black stripe down the back ; under-parts white with a tinge of yellow. Ears somewhat variable in size, nearly naked internally, and at their tips outside, followed posteriorly by a white or pale yellow spot. Upper surface of parachute dark brown or greyish, the edges fringed with white or pale yellow; feet pale grey or white on their upper surfaces. Length of toes in the order 5, 4, 3, 2, 1. Tail moderate, very bushy, grey, darkening terminally into black. Length of head and body about 10 inches ; of tail 11 inches.

Distribution.—Eastern Australia, from Queensland to Victoria.

II. LESSER FLYING PHALANGER. PETAURUS BREVICEPS.

Petaurus (*Belideus*) *breviceps*, Waterhouse, Proc. Zool. Soc., 1838, p. 152.
Petaurus breviceps, Gray, in Grey's Australia, Appendix, vol. ii., p. 402 (1841) ; Thomas, Cat. Marsup. Brit. Mus., p, 156 (1888).
Belideus breviceps, Lesson, Nouv. Tabl. Règne Anim., Mamm., p. 189 (1842).
Belideus ariel, Gould, Proc. Zool. Soc., 1842, p. 11.
Petaurus (*Belideus*) *notatus*, Peters, Monatsber. Ak. Berlin, 1859, p. 14.

(*Plate XVI.*)

Characters.—Size small ; fur soft and silky. General coloration as in *P. sciureus*, but with the dark stripe down the back generally indistinct. Ears large. Relative lengths of fore toes in the order 4, 5, 3, 2, 1. Tail decidedly more bushy at the base than in the species mentioned. Length of head and body about 7 inches ; of tail 8 inches.

Distribution.—Northern and Eastern Australia. Introduced into Tasmania in 1835.

Variety.—In New Guinea and the adjacent islands, from the Halmahéra group eastwards to New Britain, the typical form of this species, which, except for the smaller size of its cheek-teeth, is frequently very difficult to separate from *P. sciureus*, is replaced by a very well-marked variety. In this Papuan Flying Phalanger (var. *P. papuanus*) the fur is shorter and closer than in the typical form ; while, owing to this shortness, the stripe on the back and the other markings are more distinctly defined. The relatively small ears are also generally narrower and less rounded than in the Australian form ; while there is some difference in the hairing of the under surface of the hind feet. The teeth are likewise relatively smaller and lighter.

If these differences were invariable, they might perhaps be taken to indicate that the Papuan form is a distinct species, but the specimens from the Aru Islands tend to bridge over the difference between the Australian and New Guinea forms.

III. YELLOW-BELLIED FLYING PHALANGER. PETAURUS AUSTRALIS.

Petaurus australis, Shaw, Nat. Miscell., vol. ii., pl. lx. (1791) ; Thomas, Cat. Marsup. Brit. Mus., p. 151 (1888).

Didelphys petaurus, Shaw, Gen. Zool., vol. i., pt. 2, p. 496 (1800).

Phalangista petaurus, Cuvier, Règne Animal, vol. i., p. 180 (1817).

Petaurus flaviventer, Desmarest, **Nouv.** Dict. d'Hist. Nat., vol. xxv., p. 403 (1817).

Belideus flaviventer, Lesson, Nouv. Tabl. Règne Anim., Mamm., p. 189 (1842).
Belideus flaviventris, Bennett, Gatherings of a Naturalist, p. 150 (1860).

Characters.—Size large; fur long; nose large, naked, finely granulated. General colour brown, variously marked with orange and black; chin and inner sides of wrists and ankles blackish; remainder of under-parts deep orange. Ears long, narrow, naked internally, and at the tip externally, with a prominent yellow patch along the posterior margin. Middle line of back and upper surfaces of parachute dark brown; edges of parachute orange, except at its origin and insertion, where it has a broad black fringe; feet black on upper surface. Length of fore toes in the order 5, 4, 3, 2, 1. Tail very long, bushy, grey above, inferiorly orange, darkening to black at the tip. Length of head and body about 12 inches; of tail 17 inches.

Distribution.—Coast ranges of New South Wales and Victoria.

Habits.—The following excellent account of the habits of this species is given by Bennett, who had a young example in his possession, which he succeeded in raising to maturity:—

"The animal," this naturalist writes, "from the conformation of its feet, is evidently intended to live in trees, and therefore, when seen on the ground, has a very awkward waddling gait. This is shown but seldom, and then only when it is obliged to walk upon the level surface. When climbing up a tree it becomes more independent in character, and it regards the spectator from the top of its perch in a very different manner. It retires either between the forked branches or in the hollow cavities of the tree during the day to sleep, and at night passes from one to another by flying leaps, aided by its parachute-like membrane, descending to the ground only from unavoidable necessity, such as when the trees are so far apart as to render it impossible to traverse the space by leaping. When pursued, it takes to the highest branches, and springs from tree to tree with great rapidity, reminding me of monkeys I have seen in the forests of Singapore, which, when frightened, exhibit a similar degree of activity. It contrives to elude its pursuers by leaps, which, giving an impetus to the body, are very materially aided by the expanded membrane between the fore and hind feet. This enables the animal to pass over a very considerable distance in its leaps. It is surprising to see it jumping from branch to branch and tree to tree, in the clear and delightful atmosphere of a fine Australian moonlight night, with so extraordinary a degree of skill and rapidity. But I remarked that the flying leaps were invariably downwards, in an oblique direction, and, that when desirous of ascending, the creature would climb rapidly, and if overtaken would cling so tenaciously to the bark of the tree, as, while living, to be very difficult of removal. Having become tamer from confinement, the animal would suffer itself to be handled without scratching and biting as at first, and would lick the hand for sweets, of which it was very fond, and permit its little nose to be touched and fur examined in any gentle manner; but if anyone attempted to take it up by the body it became most violent in temper, biting and scratching with savage rage, at the same time uttering its snarling, wheezing, spitting kind of guttural growl. If caught by the tail it would be more quiet (excepting if held too long in one position), and would spread the membrane as if to save itself from falling. . . . It is a crepuscular and night animal, sleeping most of the day coiled up in a circle, with its bushy tail thrown over it like a blanket; it occasionally wakes up and feeds a little, but appears then to be defective in vision and unable to endure the strong glare of daylight, soon seeking its dark retreat and repose; but in the dusk of evening and at night it is in full life and activity—not the dull, lifeless animal seen during the day. When in its cage, it turns over and over the perch, is very restless, climbs up on the bars, and is in incessant action; when set at liberty it mounts to the highest part of any object in the room, and seems then quite independent, and in a happy and contented state of mind.

"It was fed upon milk, raisins, and almonds; indeed, sweets of all kinds, in the form of preserved fruits especially, as well as loaf-sugar, met with its approbation; it licked the sweets, and devoured the pulp only of the preserved fruits, leaving the skins. It appeared a very small eater, but fattened and throve well. In its wild state it feeds upon the honey of the eucalypti or gum-trees, as well as on the tender shoots and seeds. No doubt insects form a portion of its diet."

After certain observations regarding the pelage and general appearance of these pretty little animals, the writer goes on to say that "the blacks capture them for food, and having prepared them by singeing the fur, cook them with the skins on, which gives the meat a more delicate and juicy flavour; but by the colonists they are valued only for their fur, which, for delicacy and beauty, almost equals that of the chinchilla. This animal traverses the tops of the trees, and passes to the extremity of the outermost branches with the greatest facility. When leaping, it is observed always to ascend a little at the termination of the leap, by which the shock received in coming from a great height is broken.

"My captured specimen escaped one night from its place of confinement, and was seen in one of the uppermost branches of a lofty weeping-willow tree, quietly reposing between one of the forks of the larger branches. A boy was sent to climb up the tree to come upon the animal when asleep. By creeping up cautiously he approached the creature without being seen or heard, and succeeded in seizing it by the tail, threw it down a height of about sixty feet, when, by the assistance of its parachute-like membrane, it alighted safely upon the ground, and was then readily secured again. It holds a raisin or almond in its fore paws, licking and nibbling it. It is often seen lying on its back at the bottom of the cage when feeding, and when drinking milk holds the small vessel containing it between its fore paws, lapping like a kitten. It is evident, from the fondness of this animal for sweets, that, when the eucalypti are in flower, it subsists upon honey, which the blossoms yield in very large quantity (the honey is in such abundance as to afford subsistence to honey-eating Parrots and other birds, as well as to these animals, and also to myriads of insects of various species). When these have disappeared, it lives upon the nuts and young foliage, and probably, as is usual with honey-feeding animals, also upon insects. It drinks frequently, and will take water, but evinces a decided preference for, and thrives best upon, milk. I found that it would sometimes eat the young flower-buds of the eucalyptus, and was also fond of succulent fruit, such as apricots."

According to a later observer, this Flying Phalanger feeds on moths, leaves, and berries.

In reference to the foregoing statement that these animals are hunted by the colonists for the sake of their skins, it may be mentioned, on the authority of Mr. Poland, that the fur, like that of all flying mammals, is unsuited to the purposes of the manufacturer, in consequence of the extreme tenuity of the skin.

THE SQUIRREL-PHALANGERS. GENUS GYMNOBELIDEUS.

Gymnobelideus, McCoy, Ann. Mag. Nat. Hist., ser. 3, vol. xx., p. 287 (1867).

Size small; general form and appearance as in *Petaurus*, save for the absence of a parachute-like expansion of the skin of the flanks. Ears large, naked, without tufts. Toes of normal proportions, their relative lengths being in the order 4, 3, 5, 2, 1; claws shorter than in *Petaurus;* tail long, cylindrical, and bushy. Skull and teeth as in the genus last-named.

The rare and little-known animal which is the sole representative of this genus appears to be closely allied to the ancestral form from which the Flying Phalangers of the genus *Petaurus* have originated.

I. LEADBEATER'S PHALANGER. GYMNOBELIDEUS LEADBEATERI.

Gymnobelideus leadbeateri, McCoy, Ann. Mag. Nat. Hist., ser. 3, vol. xx., p. 287 (1867); Thomas, Cat. Marsup. Brit. Mus., p. 149 (1888).

Characters.—Fur soft and close. General colour brownish-grey; under-parts yellowish; a dark median streak along the nape and back; a dark patch below the base of the ear, and fainter ones above and below the eye. Ears large, semi-elliptical, and nearly naked towards the tips; toe-pads of fore feet large and wrinkled; hinder pads of both fore and hind feet large, low, and finely striated. Tail pale brown. Length of head and body about $5\frac{1}{2}$ inches; of tail $6\frac{1}{2}$ inches.

Distribution.—Victoria, in the neighbourhood of the Bass river.

THE DORMOUSE-PHALANGERS. GENUS DROMICIA.

Dromicia, Gray, in Grey's Australia, Appendix, vol. ii., p. 407 (1841).

Size small; ears large, thin, and almost naked; no parachute-like expansion on the flanks; toes normally proportioned, the relative lengths of those of the fore limb being 3, 4, 2, 5, 1; fore claws short and rudimentary, hind ones normal. Tail cylindrical, furry at the base, elsewhere scaly and clothed with short hairs at the extreme tip roughened and naked inferiorly and prehensile. Molar teeth evenly rounded, with four cusps, with the exception of the last, which is frequently absent.

This genus, which includes four species, is distributed over New Guinea, Western Australia, and Tasmania. "It is," remarks Mr. Thomas, "evidently intermediate between *Acrobates* and *Petaurus*, and has had apparently to give way to these more highly specialised and, presumably, later forms wherever the two have come in contact. Of this the distribution of the genus is a curious example, since it is isolated in the three places most conspicuous for their retention of ancient forms— New Guinea, Western Australia, and Tasmania—while no species appears now to live in the temperate parts of Eastern Australia, where the more highly developed genera above referred to have their principal home, and where, judging by its distribution, *Dromicia* must obviously at one time have lived."

1. WESTERN DORMOUSE-PHALANGER. DROMICIA CONCINNA.

Dromicia concinna, Gould, Proc. Zool. Soc., 1845, p. 2; Thomas, Cat. Marsup. Brit. Mus., p. 146 (1888).
Phalangista (Dromicia) concinna, Waterhouse, Nat. Hist. Mamm., vol. i., p. 314 (1846).
Phalangista (Dromicia) neilli, Waterhouse, *op. cit.,* p. 315.

Characters.—Size small; form very light and delicate. General colour bright fawn; hairs of under-parts pure white throughout; dark eye-mark nearly obsolete. Ears long, rather narrow, evenly oval. Limbs fawn-coloured externally, white on the inner side. Tail slender. Only three pairs of molar teeth. Length of head about 3 inches; of tail slightly more.

This species may be distinguished from all the other three by the hairs of the under-parts being pure white throughout their length, instead of slate-coloured at the roots; and also by

PLATE XVII.

COMMON DORMOUSE-PHALANGER

the minute size of the fourth premolar, which is not larger than the other teeth of the same series.

Distribution.—South and West Australia.

Habits.—In size rather inferior to the common English Dormouse, this beautiful little creature is abundantly and generally distributed over the Swan river district. Being strictly nocturnal in its habits, it secretes itself, according to Gould's account, during the day in the hollows of trees, and at night leaves its retreat for the flowering branches of low shrubby trees. At night it is very active, and when in confinement will then leap across its cage in pursuit of insects. Another observer states that it is found under the dead bark of eucalpytus and other trees, and in holes in trees that have been excavated by fire. In such situations, if due caution be observed, it may readily be captured by the hand.

II. COMMON DORMOUSE-PHALANGER. DROMICIA NANA.

Phalangista nana, Desmarest, Nouv. Dict. d'Hist. Nat., vol. xxv., p. 477 (1817).
Phalangista gliriformis, Bell, Trans. Linn. Soc., vol. xvi., p. 121 (1828).
Dromicia nana, Gray, in Grey's Australia, Appendix, vol. ii., p. 401 (1841); Thomas, Cat. Marsup. Brit. Mus., p. 144 (1888).
Dromicia unicolor, Krefft, Proc. Zool., 1863, p. 49.

(*Plate XVII.*)

Characters.—Size large; form rather thick and clumsy; fur thick, soft, and somewhat woolly; nose naked, finely granulated. General colour uniform dull fawn; under-parts slaty, with the tips of the hairs white; dark eye-mark indistinct. Ears large, narrow, evenly **oval**. Limbs grey, with the fore feet brown, and the hinder ones whitish. Tail rather long, with the basal inch thickened. Length of head and body about 4 inches; of tail rather more.

Distribution.—Tasmania.

Habits.—This elegant little Phalanger, according to Gould, is more especially abundant in the northern districts of Tasmania than elsewhere in the country; and of all trees it apparently prefers the Banksias, whose numerous blossoms supply it with an unfailing supply of honey and insects. During the day it generally lies coiled up asleep in some crevice or hollow branch.

During the winter months it becomes less active, and undergoes a kind of hibernation, although less complete than in the case of the Dormouse.

Writing from observations made on specimens kept in captivity, Bell observes that in their habits these little Phalangers "are extremely like the Dormouse, feeding on nuts and other similar food, which they hold in their fore paws, using them as hands. They are nocturnal, remaining asleep during the whole day, or, if disturbed, not easily roused to a state of activity; and coming forth late in the evening, and then assuming their natural rapid and vivacious habits, they run about a small tree which is placed in their cage, using their paws to hold by the branches, and assisting themselves by their prehensile tail, which is always held in readiness to support them, especially when in a descending attitude. Sometimes the tail is thrown in a reverse direction, turned over the back, and at other times, when the weather is cold, it is rolled closely up towards the under-part, and coiled up almost between the thighs. When eating, they sit upon the hind quarters, holding the food in their fore paws, which, with the face, are the only parts apparently standing out from the ball of fur, of which the body seems at that time to be composed. They are perfectly harmless and tame, permitting anyone to hold and caress them, without even attempting to bite, but do not evince the least attachment to persons about them or even to each other."

III. LONG-TAILED DORMOUSE-PHALANGER. DROMICIA CAUDATA.

Dromicia caudata, Milne-Edwards, Comptes Rendus, vol. lxxxv., p. 1079 (1879); Thomas, Cat. Marsup. Brit. Mus., p. 143 (1888).

Characters.—The largest species; fur very long, soft, and silky. General colour dull rufous; under-parts pale yellowish-white, with the bases of the hairs slaty; face rufous fawn, with two broad black bands passing from the nose through the eyes nearly to the ears. Ears rather long and narrow, regularly oval, naked; legs grey; feet dull fawn; tail much longer than the head and body, very thin, furry like the body for half-an-inch at the root, but elsewhere short-haired. Four molar teeth.

Distribution.—North-west New Guinea.

IV. LESSER DORMOUSE-PHALANGER. DROMICIA LEPIDA.

Dromicia lepida, Thomas, Cat. Marsup. Brit. Mus., p. 142 (1888).

Characters.—Size small; form slender and graceful; fur fine, soft, and silky; nose finely granulated. General colour pale bright fawn; hairs of under-parts dark slaty tipped with white; a conspicuous dark eye-mark. Ears large, broad, and nearly naked; soles of feet naked and finely granulated; tail rather long, with the basal half-inch thickly furred, the remainder short-haired and slightly scaly. Four pairs of molar teeth. Length of head and body rather less than that of the tail, which is about 3 inches.

Distribution.—Tasmania.

For a long time this well-marked species was mistaken for the young of *Dromicia nana*, until its distinctive features were pointed out by Mr. Thomas. Although externally the two are very similar, they present marked points of difference in the skull and teeth.

THE PEN-TAILED PHALANGERS. GENUS DISTÆCHURUS.

Distæchurus, Peters, Ann. Mus. Genova, vol. vi., p. 303 (1874).

Size small; ears rather short, thinly haired, but with small tufts round the base; no parachute-like expansion of skin on the flanks; toes of normal proportions; claws sharp and curved; tail with the long hairs arranged in two opposite lateral rows like the vanes of a feather. Molar teeth three in number, small and rounded, with smooth unridged cusps; the last premolar very small in the upper, and wanting in the lower jaw. Two teats.

PEN-TAILED PHALANGER. DISTÆCHURUS PENNATUS.

Phalangista (Distæchurus) pennata, Peters, Ann. Mus. Genova, vol. vi., p. 303 (1874).
Phalangista pennata, Ramsay, Proc. Linn. Soc., N. South Wales, vol. ii., p. 12 (1878).
Distæchurus pennatus, Thomas, Cat. Marsup. Brit. Mus., p. 139 (1888).

Characters.—Form very Dormouse-like; fur soft, thick, and woolly. General coloration of head striped, of body dull buff; under-parts white; face white, with two dark bands running from the sides of the muzzle through the eyes to between the ears. Naked portion of muzzle sharply defined, and pentagonal in shape. Outer sides of limbs like back, inner sides like under-parts; soles of feet smooth, with low and rounded pads; tail considerably longer than the head and body, the basal half-inch thickly furred, and the remainder naked, save for the lateral fringes of long hairs. Length of head and body nearly as in *Dromicia nana*, but the tail about half as long again.

Distribution.—New Guinea.

Although in size, coloration, and general appearance very different, this little Phalanger is, as Mr. Thomas remarks, very closely allied to the undermentioned Pigmy Flying Phalanger as regards its skull and teeth, and may, indeed, be very close to the ancestral form from which the latter was evolved. Although far less specialised than its Australian ally, the Papuan form, as in so many analogous instances, is the more brilliantly coloured of the two animals.

THE PIGMY FLYING PHALANGERS. GENUS ACROBATES.

Acrobates, Desmarest, Nouv. Dict. d'Hist. Nat., vol. xxv., p. 105 (1817).

Size very small; ears medium; flanks with a narrow parachute-like expansion of skin; toes of normal proportions, each provided with a broad, striated, terminal pad; lengths of those of the fore foot in the order 4, 3, 5, 2, 1; claws sharp and well-developed, although not very prominent; tail as in *Distœchurus*. Teeth as in the latter, except that the last premolar is larger, and present in both jaws. Teats four.

This genus is represented only by two species, one of which is Australian and the other Papuan.

I. PIGMY FLYING PHALANGER. ACROBATES PYGMÆUS.

Didelphis pygmæa, Shaw, Zool. New Holland, vol. i., p. 5 (1794).
Phalangista pygmæa, Geoffr., Cat. Mus., p. 151 (1803).

Acrobates pygmæus, Desmarest, Nouv. Dict. d'Hist. Nat., vol. xxv., p. 405 (1817); Thomas, Cat. Marsup. Brit. Mus., p. 136 (1888).
Petaurus pygmæus, Lesson, Dict. Class. d'Hist. Nat., vol. xiii., p. 289 (1828).
Petaurus (*Acrobates*) *pygmæus*, Waterhouse, in Jardine's Naturalist's Library, Mamm., vol. xi., p. 293 (1841).
Dromicia frontalis, De Vis, Proc. Linn. Soc., N. South Wales, ser. 2, vol. i., p. 1134 (1887).

(*Plate XVIII.*)

Characters.—Form very light and delicate; fur soft, straight, and silky; a well-defined naked area on the muzzle. General colour greyish-brown, the under-parts and inner sides of limbs white; area round and in front of eyes brown; feet brown. Tufts of hairs present behind the eyes and inside the ears, the latter being fawn-colour on the outer sides anteriorly and white posteriorly; margins of parachute fringed with longish hairs. Tail rather long, fawn-colour, with its extreme tip naked inferiorly and probably prehensile. Length of head and body about 3 inches; of tail nearly the same.

Distribution.—Queensland, to the south of latitude 20°, New South Wales, and Victoria.

Habits.—Resembling a Common Mouse in size, and hence known to the colonists as the Flying Mouse, or Opossum-Mouse, this little animal is one of the most elegant of the Australian Marsupials. At one time exceedingly numerous in the neighbourhood of Port Jackson, although comparatively rare in other parts of its habitat, it is, from its small size, but seldom seen, although individuals will at times come into the tents of those camping out in the bush. Beyond the fact that it is arboreal and volant, little seems to have been recorded of its

PLATE X

PIGMY FLYING-PHALANGER.

habits; but an anonymous writer states that he has seen a family of young ones taken out of a hollow tree.

II. PAPUAN PIGMY FLYING PHALANGER. ACROBATES PULCHELLUS.

Acrobates pulchellus, Rothschild, Proc. Zool. Soc., 1892, p. 546.

Characters.—According to its describer, this species differs from the last in its more purplish-brown colour, its broader and more robust head, much shorter tail, and comparatively smaller body. The under-parts are also much whiter, and the whole of the throat and sides of the lower jaw are pure white, instead of yellowish-grey. Around the eyes is a blackish-brown patch extending nearly to the nose, while the ears are rather smaller than in *A. pygmæus*. The length of the tail is $2\frac{1}{2}$ inches against $3\frac{1}{2}$ in the latter.

Distribution.—One of the small islands of Northern Dutch New Guinea.

LONG-SNOUTED PHALANGERS. GENUS TARSIPES.

Tarsipes, Gervais and Verreaux, Proc. Zool. Soc., 1842, p. 1.

The last group of the Phalangers represented only by a single small species, differs so markedly from all the rest that it constitutes a sub-family—the *Tarsipedinæ*—by itself; its characters being as follows: Tail long; muzzle very long and slender; tongue extensile; intestine without a blind appendage or cæcum; cheek-teeth minute and rudimental.

The foregoing characters serve to distinguish the group as a sub-family, and the following may be taken as characterising the one known genus: Size small; form slender; head long and narrow; ears of moderate size, thinly haired; soles of feet naked and granulated. Claws rudimentary, except those of the united second and third toes of the hind foot; tail long, thinly haired, and prehensile. Four teats. Upper canine and lower incisor teeth comparatively well-developed; at most but three pairs of molar teeth, and the premolars represented only by the last of the series in the upper jaw. The slender lower jaw is remarkable among Marsupials for the absence of any inflection of its hinder angle—a feature doubtless due to extreme specialisation and the general feebleness of the skull and jaws.

THE LONG-SNOUTED PHALANGER. TARSIPES ROSTRATUS.

Tarsipes rostratus, Gervais and Verreaux, Proc. Zool. Soc., 1842, p. 1; Thomas, Cat. Marsup. Brit. Mus., p. 133 (1888).

Tarsipes spenseræ, Gray, Ann. Mag. Nat. Hist., vol. i., p. 40 (1842).

Characters.—Fur short, coarse, and rough; naked area of nose sharply defined, and finely granulated. General colour grey, striped with black or brown; under-parts yellowish-white; an indistinct pale area round each eye; legs grey; feet white; ears rounded; soles of feet with five pads; fourth and fifth toes of hind foot disproportionately long, and almost clawless, like the first toe; second and third toes of hind foot very completely united. Tail brown above, white or pale yellow on the sides and below, with the extreme tip naked. Length of head and body about 3 inches; of tail 4 inches.

Distribution.—West Australia.

Habits.—The Long-snouted Phalanger, which derives its scientific name from a certain resemblance of its hind feet to those of a Malayan Lemur-like animal known as the Tarsier, is one of the most interesting members of the Phalangers, as showing the modifications of form and structure produced by specialisation for a particular mode of life. Known to the natives by the name of *Tait*, and *Nulbenger*, the Long-snouted Phalanger, writes Gould, "is generally found in all situations, suited to its existence, from Swan River to King George's Sound. He adds · "From its rarity, and the difficulty with which it is procured, notwithstanding the high rewards offered, the natives only brought me four specimens; one of these, a female, I kept alive for several months, and it became so tame as to allow itself to be caressed in the hand without evincing any fear, or making any attempt to escape. It was strictly nocturnal, sleeping during the greater part of the day, and becoming exceedingly active at night. When intent upon catching flies, it would sit quietly in one corner of its cage, eagerly watching their movements, as, attracted by the sugar, they flew around; and when a fly was fairly within its reach, it bounded as quick as lightning, and seized it with unerring aim, then retired to the bottom of the cage, and devoured it at leisure, sitting tolerably erect and holding the fly between its fore paws, and always rejecting the head, wings, and legs. The artificial food given it was sopped bread, made very sweet with sugar, into which it inserted its long tongue, precisely in the way in which the honey-eaters among birds do theirs into the flower-cups for honey. Every morning the sop was completely honeycombed, as it were, from the repeated insertion of the tongue; a little moistened sugar on the end of the finger would attract it from one part of the cage to the other; and by this means an opportunity may be readily obtained for observing the beautiful prehensile structure of the tongue, which I have frequently seen protruded for nearly an inch beyond the nose. The edges of the tongue near the tip are slightly serrated. The tail is prehensile, and is used when the animal is climbing, precisely like that of the *Hepoona* (*Pseudochirus*). The eyes, although small, are exceedingly prominent, and placed very near to each other; the ears are generally quite erect. When sleeping, the animal rests upon the lower part of the back, with its long nose bent down between its fore feet, and its tail brought over all, and turned down the back. Mr. Johnson Drummond shot a pair in the act of sucking the honey from the blossoms of the *Melaleuca*; he watched them closely, and distinctly saw them insert their long tongues into the flower precisely after the manner of the birds above mentioned."

Another observer—Neill—states that in the neighbourhood of King George's Sound the Long-snouted Phalanger makes its nest in the overhanging foliage of the kingias and other large-leaved plants. In several examples described by him the stomach contained only a small quantity of clear honey-like fluid, thus confirming the statements of the natives that the animal in its wild state lives almost entirely by thrusting its extensile tongue into the tubes of flowers for the sake of extracting their honey. Since nearly all the Australian flowers are honey-yielding, the creature would have no difficulty in obtaining nutriment in this manner throughout the year.

THE WOMBATS. FAMILY PHASCOLOMYIDÆ.

The third and last family of the Diprodont Marsupials is represented solely by the Wombats of Australia and Tasmania, all of which are included within the limits of a single genus. Heavily-made and short-limbed creatures, with incisor teeth curiously resembling those of the Rodent Mammals, the Wombats may be regarded as filling in Australia the place occupied

THE WOMBATS.

in the northern hemisphere by the Marmots and in South America by the Viscachas, both of which are members of the Rodentia. In this instance, indeed, the resemblance is not confined to similarity of habits, since the Wombats not only resemble Marmots (save for their rudimentary tails), but also simulate them to a great extent in the structure and arrangement of their teeth, more especially in the form and number of their incisors, and in the total absence of canines. They thus afford an excellent instance of that parallelism in development which we have already alluded to as occurring among totally

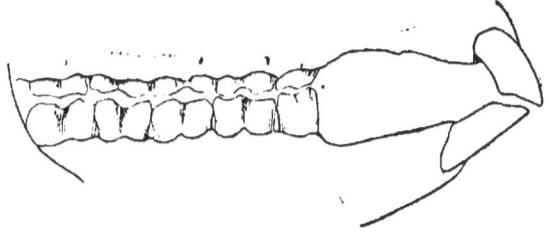

Side View of the Teeth of Wombat.

different groups of animals, in the case of those adapted to similar modes of life and living under similar surroundings. Such a parallelism is, however, wanting in the case of the Kangaroos and Wallabies, which may be regarded as representing in the Australian economy of nature the Ruminants of the Old World, since there is no similarity in the structure of the members of the two groups. The same holds good with regard to the Phalangers, which may be regarded as the representatives in Australia of both the Squirrels, Lemurs, and Monkeys of other regions. That the Flying Phalangers, which represent the Flying Squirrels of the Oriental Region and Africa, resemble the latter in having a parachute-like expansion of skin along the flanks, is indeed true; but this can scarcely be regarded as an instance of parallelism in its proper sense, since it is obvious that without the special development of the fore limbs characterising the Bats, there is no method by which Mammals could take flying leaps save by such a development of the skin on the sides of the body.

As a family, the Wombats are distinguished by the following combination of characters:—

Form stout and clumsy; muzzle short and broad; limbs subequal, thick, short, and strong; fore feet with five subequal toes, each furnished with a powerful claw; in the hind feet the first toe, or hallux, short and clawless, the remainder with long, curved claws, the second and third of the series being imperfectly united in a common membrane. Tail rudimental; stomach simple; intestine with a blind appendage, or cæcum. Teeth growing continuously, and thus never forming roots; a single pair of incisors in each jaw, which are large, curved, strong, and chisel-like, having enamel only on their front and lateral surfaces; no canines. Cheek-teeth five in number, and separated by a long gap from the incisors; the molars strongly curved, with two lobes, the convexity being internal in those of the upper, and in the opposite direction in those of the lower jaw; premolar with only a single lobe, but otherwise similar to the molars.

The Wombats are confined to Australia and Tasmania south of the tropics; and in habits are digging and root-eating animals.

WOMBATS. GENUS PHASCOLOMYS.

Phascolomys, Geoffr., Ann. Muséum, vol ii., p. 364 (1803).

The three species of Wombat are included in this genus,

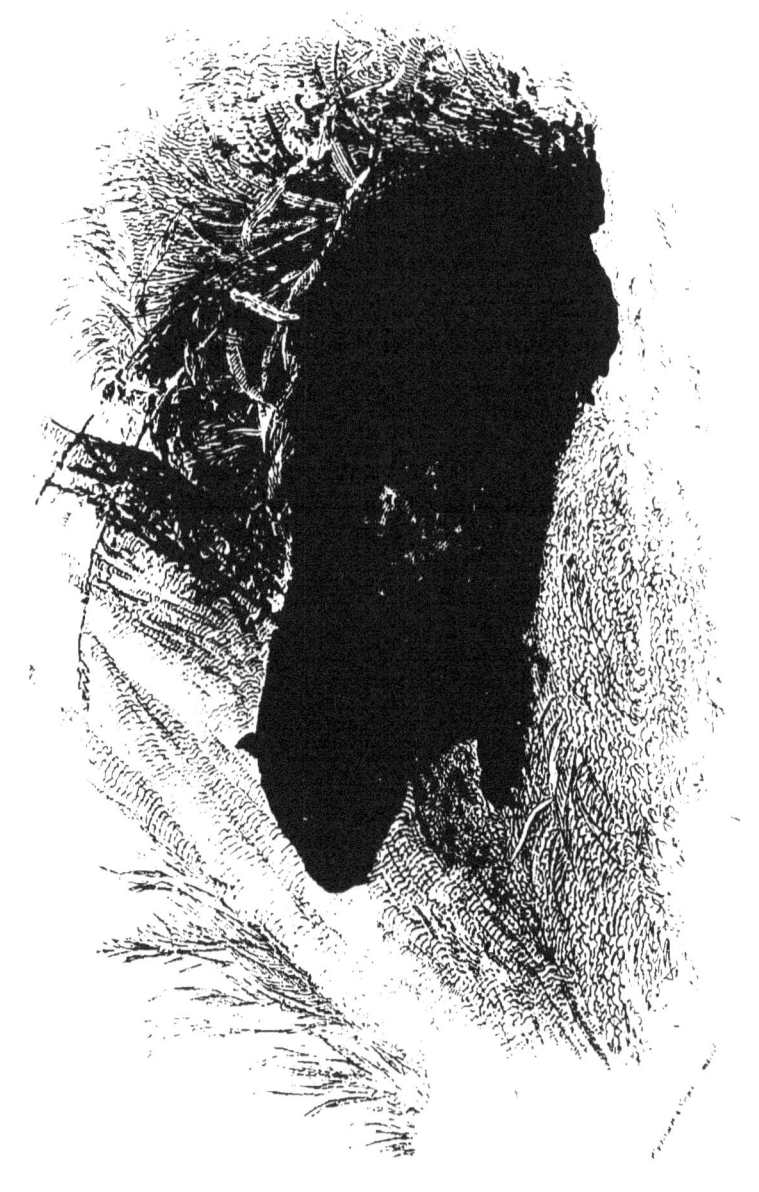

and their characters may be taken as identical with those of the family.

I. COMMON WOMBAT. PHASCOLOMYS MITCHELLI.

Phascolomys mitchelli, Owen, in Mitchell's Exped. Australia, vol. ii., p. 362 (1838); Thomas, Cat. Marsup. Brit. Mus., p. 213 (1888).

Phascolomys platyrhinus, Owen, Cat. Osteol. Mus. R. Coll. Surg., vol. i., p. 334 (1853).

Characters.—Size large; naked portion of muzzle large; fur coarse, harsh, and rough. Colour either yellow, grizzled yellow and black, or black. Ears short rounded, and fully haired. Fifteen pairs of ribs. Length of head and body about 44 inches.

Distribution.—New South Wales, Victoria, and South Australia.

Varying considerably in coloration, but apparently only individually, and not according to geographical distribution, this species is noteworthy as having been originally described on the evidence of fossil remains from the superficial deposits of Australia. Many years, however, elapsed before the so-called *Phascolomys platyrhinus* was ascertained to be inseparable from the fossil form. Its habits may be considered in connection with those of the next species.

II. TASMANIAN WOMBAT. PHASCOLOMYS URSINUS.

Didelphis ursina, Shaw, Gen. Zool., vol. i., pt. ii., p. 504 (1800).

Wombatus fossor, Desmarest, Nouv. Dict d'Hist. Nat., vol. xxiv., p. 20 (1803).

Phascolomys fusca, Illiger, Prodz. Syst. Mamm., p. 78 (1811).
Phascolomys vombatus, Leach, Zool. Miscell., vol. ii., p. 102 (1815).
Phascolomys wombat, Péron and Lessueur, Voyage Terr. Austral., vol. ii., p. 13 (1816).
Phascolomys ursinus, Cuvier, Régne Animal, vol. i., p. 185 (1817); Thomas, Cat. Marsup. Brit. Mus., p. 215 (1888).
Phascolomys bassii, Lesson, Man. Mamm., p. 229 (1827).
Phascolomys fossor, Wagner, in Schreber's Säugeth., Suppl., vol. iii., p. 132 (1843).

(*Plate XIX.*)

Characters.—Differs from the preceding species merely by being about one-fourth less in size. Colour uniform dark grizzled greyish-brown. Length of head and body about 38 inches.

Distribution.—Tasmania, and the islands of Bass Straits.

In remarking on its inferior size, as compared with its cousin of the mainland, Mr. Thomas observes that this is an instance of the reversal of the prevalent rule that the Mammals of Tasmania exceed in dimension their nearest allies inhabiting continental Australia.

Habits.—Like most Marsupials, the Wombats are essentially nocturnal animals, remaining concealed throughout the day in their subterranean quarters, from whence they issue forth at night to feed. They are the only members of the Diprotodont division of the order which are thoroughly fossorial, and it appears that they generally excavate their own dwellings in the ground, although they may take advantage of natural clefts or holes. Their food consists partly of grass and other herbage, but mainly of roots, which their powerful front teeth are admirably adapted to gnaw. From specimens kept in confinement, it appears that the female usually produces from three to four young at a birth, which are tended with great care and solicitude, until such time as they are able to shift for themselves. By no means active in their movements, and shuffling along with an awkward gait which calls to mind the progression of a young Bear, the Wombats are gentle and somewhat stupid in disposition, even at times allowing themselves to be captured and carried off without making the least attempt at resistance. If, however, their anger be once roused, they can bite fiercely, and inflict severe wounds. Although both the Common Wombat and the Tasmanian species utter no sound but a kind of hiss, the next species is said to give vent to a kind of groan.

Writing of the habits in captivity of one of the first examples brought to England, Sir Everard Home observes that the animal " burrowed in the ground whenever it had an opportunity, and covered itself in the earth with surprising quickness ; it was very quiet during the day, but constantly in motion in the night ; was very sensible of cold ; ate all kinds of vegetables, but was particularly fond of new hay, which it ate stalk by stalk, taking it into its mouth like a Beaver, by small bits at a time. It was not wanting in intelligence, and appeared attached to those to whom it was accustomed, and who were kind to it. When it saw them, it would put up its fore paws on their knees, and when taken up would sleep in the lap. It allowed children to pull and carry it about, and when it bit them, it did not appear to do it in anger, or with violence." It may be added that although this account was written as far back as 1808, it does not appear that fresh observations have been recorded.

Although in appearance the flesh of the Wombat is red and coarse, in its fatness and flavour it has been compared to pork. From the harshness and coarseness of the fur, the skin is only available for leather.

III. HAIRY-NOSED WOMBAT. PHASCOLOMYS LATIFRONS.

Phascolomys latifrons, Owen, Proc. Zool. Soc., 1845, p. 82; Thomas, Cat. Marsup. Brit. Mus., p. 217 (1888).
Phascolomys lasiorhinus, Gould, Mamm. Austral., vol. i., pl. lix. (1863).

Characters.—Size intermediate between that of the other two species; muzzle covered with velvety white hair; fur straight, soft, and silky. General colour mottled grey; tip of muzzle, a spot above and another below each eye, cheeks, throat, and chest white; chin black; remainder of under-parts grey. Ears comparatively long, narrow, and pointed, sparsely clothed externally with black hairs, internally naked. Thirteen pairs of ribs. Length of head and body about 40 inches.

Distribution.—South Australia.

THE BANDICOOTS. FAMILY PERAMELIDÆ.

With the Wombats we take leave of the Diprotodont, or first primary subdivision of the Marsupials, and the Bandicoots bring us to the first family of the second or Polyprotodon sub-division of the order, which includes the whole of its remaining representatives.

The Polyprotodonts derive their title from the presence of a numerous series of incisor teeth in the upper jaw; there being generally four or five nearly equal-sized pairs of these teeth in the upper, and three in the lower jaw, although in the case of one specialised genus the number is reduced to three pairs in both the upper and the lower jaw. In size the incisors are small, and the tusks are large and pointed; while there is never that long gap between the incisor and molar series, which forms such a characteristic feature in the dentition of many of the

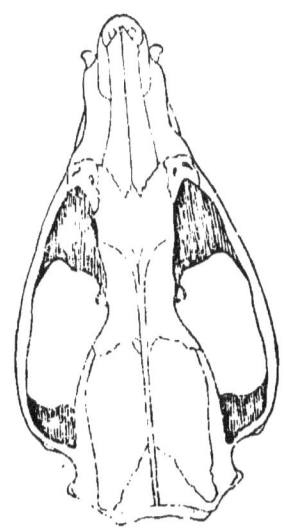
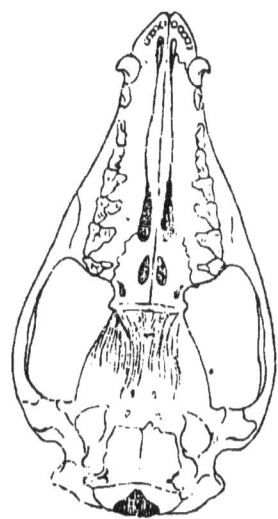

Skull and Teeth of an Opossum.

Diprotodonts. As an almost invariable rule, the molar teeth of the members of this division have their crowns surmounted by a number of sharp cusps; the general characters of the dentition being well shown in the accompanying figures of the skull of an Opossum.

From the number and characters of their teeth, it is quite evident that the Polyprotodont Marsupials are more generalised animals than their Diprotodont allies. Evidence of this is, indeed, afforded by the traces of numerous vestigial incisors occurring in fœtal Kangaroos, to which allusion has been made above; and it is confirmed by the distribution of the group, which is the only one found beyond the confines of Australia. At the present day Polyprotodont Marsupials are confined to Australasia and America; but during the Tertiary period the American Opossums were widely spread over the Old World, while large forms, apparently allied to the Thylacine, existed at the same time in South America. In still earlier epochs of the earth's history, that is to say, during the Cretaceous and Jurassic divisions of the great Secondary period, or the epochs in which our Chalk, Green-sands, and Oolites were deposited, Marsupials of this group were very widely spread over the surface of the globe, being common in Europe and North America. During these epochs, and also in the preceding Triassic epoch, when Mammals seem to have made their first appearance on the earth, Polyprotodont Marsupials, together with certain imperfectly known types, more or less intimately allied to the Monotremes, or Egg-laying Mammals, of which more anon, would appear to have been the dominant, if not the sole representatives of their class. It is also remarkable that some of these extinct European Marsupials seem to have been very closely allied to the little Banded Anteater of Australia. From the date of the Cretaceous epoch Marsupials appear to have been gradually displaced from the greater portion of the earth's surface by the more highly organised Placentals, surviving, as we have seen, only in Australia and South America, with the exception that one member of the single American family ranges into the northern half of the New World.

In habits the Polyprotodonts are chiefly carnivorous or insectivorous, and they thus take the place in Australia occupied in other regions by the true Carnivores and the so-called Insectivores, such as the Shrews, Hedgehogs, and Moles.

Although in four out of the five families into which the Polyprotodonts are divided the structure of the hind foot is quite different from that obtaining in the Diprotodonts, yet it is not a little remarkable that in the family under consideration there is the same small size and union in a common integument of the second and third toes. The relation in which this structure stands to its representative in the Diprotodonts has given rise to much discussion—but it must probably be regarded as an instance of parallel development.

With these preliminary remarks, we proceed to the special characteristics of the family *Peramelidæ*, which are as follows :—

Hind limbs markedly the longer ; fore limbs with three, or sometimes only two, of the middle toes long and clawed, the others rudimentary or absent ; hind feet with four or five very unequally-sized toes ; first toe, or hallux, rudimentary or absent; second and third slender and united in a common integument ; fourth the strongest, long, with a large claw. Pouch opening backwards ; intestine with a blind appendage or cæcum ; collar-bones, or clavicles, wanting. Four or five pairs of incisor teeth in the upper, and three in the lower jaw. Tail long and not prehensile.

This family is found both in Australia and New Guinea. Although allied to the next, it is sharply differentiated therefrom by the structure of the hind foot. The peculiarity of the hind foot was long thought to indicate affinity with the Diprotodont sub-division; but, as Mr. Thomas rightly observes, judging from the wholly Polyprotodont character of the rest of their organisation, even including the bones of the ankle, which present a much greater resemblance to those of the Dasyures than to those of the Phalangers, it seems probable that this view is erroneous, and that, as already mentioned, the syndactylous hind foot has been independently developed in the two groups. We must accordingly regard the Bandicoots as a highly specialised offshoot from the Dasyures.

In habits the Bandicoots are fossorial and insectivorous, although many subsist on a mixed diet.

THE RABBIT-BANDICOOTS. GENUS PERAGALE.

Peragalea, Gray, in Grey's Australia, Appendix, vol. ii., p. 401 (1841).

Form light and delicate; muzzle long and narrow; ears very long; fore feet with the first and third toes rudimentary and clawless, and the three middle ones long and furnished with strong curved claws; no trace of the first toe (hallux) of the hind foot externally; hind limbs much longer than the front pair; soles of hind feet hairy; tail long, and distinctly crested on the upper surface of the terminal half. Five pairs of upper and three of lower incisor teeth; molars quadrangular or rounded in section, but differing markedly in structure in the two species of the genus.

The Rabbit-Bandicoots are confined to Australia, exclusive of Tasmania, and are omnivorous in their diet.

1. COMMON RABBIT-BANDICOOT. PERAGALE LAGOTIS.

Perameles (*Macrotis*) *lagotis*, Reid, Proc. Zool. Soc., 1836, p. 129.

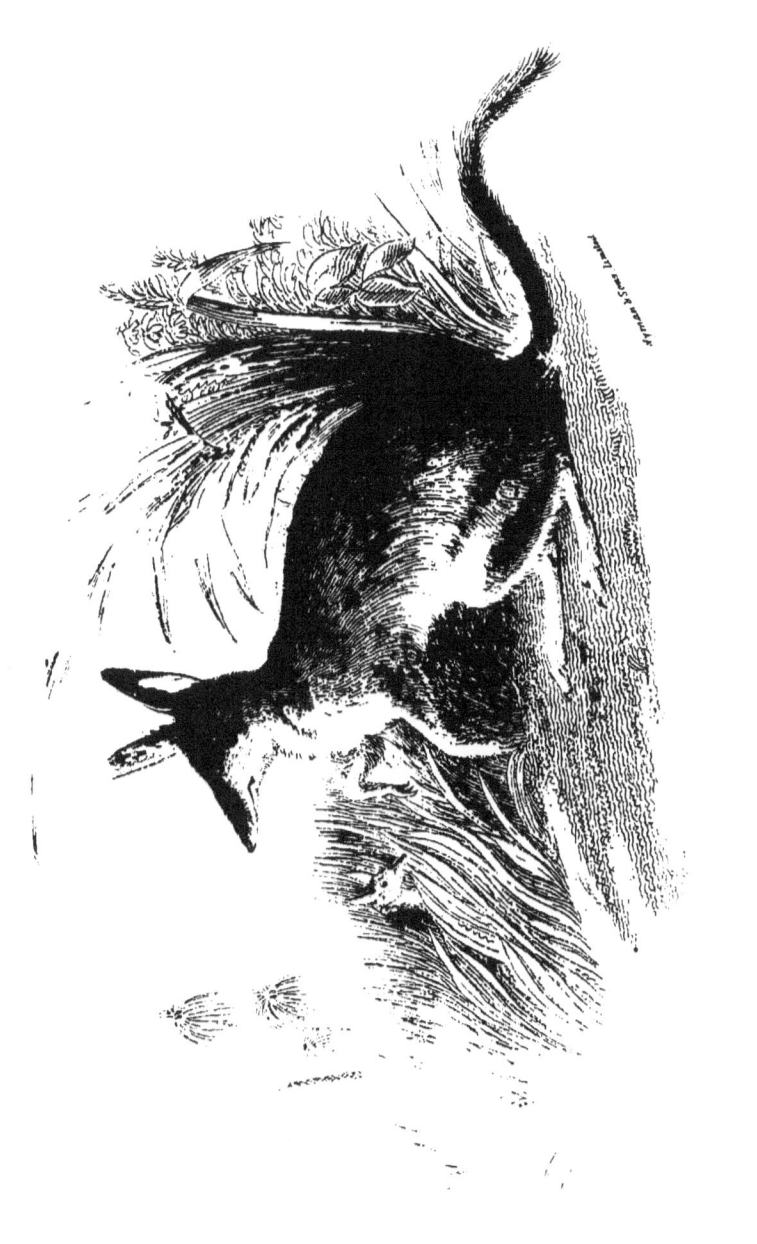

Perameles lagotis, Waterhouse, Cat. Mamm. Mus. Zool. Soc., p. 65 (1838).
Perameles (Peragalea) macrotis, Gray, in Grey's Australia, Appendix, vol. ii., p. 401 (1841).
Peragale lagotis, Gray, List Mamm. Brit. Mus., p. 96 (1843); Thomas, Cat. Marsup. Brit. Mus., p. 223 (1888).

(*Plate XX.*)

Characters.—Size large; form light and delicate; fur very long, soft, and silky. General colour fawn-grey; under-parts white; cheeks and bases of ears white or pale fawn; an indistinct darker vertical band on the side of the rump. Ears nearly naked, their edges and the anterior part of the backs thinly clothed with brown hairs. Outer sides of the fore, and backs of the hind limbs dark grey grizzled with white; elsewhere both limbs white; soles of hind feet almost wholly covered with thick hair. Tail of moderate length, thickly haired throughout, the basal third coloured like the body, the middle third black or dark brown, and the terminal moiety white and prominently crested above. Length of head and body about 18 inches; of tail 9 inches.

Distribution.—South and West Australia.

Habits.—About equal in size to an ordinary Rabbit, the common Rabbit-Bandicoot derives its colonial name of Native Rabbit, from its large and elongated ears, and consequent general resemblance to the familiar Rodent. It is stated to be fairly abundant on the extensive grassy tracts bordering the Swan river, where it is usually found in pairs. Here it usually frequents such spots as have a loose soil, suitable for the excavation of its burrows. To these holes it at once flies when pursued, and thus once more recalls the familiar European animal from which it derives the first half of its name.

The burrow has only a single entrance, and may be at once distinguished from that of Lesueur's Rat-Kangaroo. Krefft writes that as the Rabbit-Bandicoot " often prefers entering the ground on a hillside, and as hills, even of very slight elevation, are often scarce on these extensive plains [of the Murray], it will sometimes happen that the *Peragale* takes advantage of the mound raised upon a departed black-fellow's grave, providing for itself a habitation beneath the native's weary bones. Upon this ground an investigator asserted some years ago that this animal dug out the dead bodies of the natives and fed upon them." Such a charge is, however, totally unfounded, the Rabbit-Bandicoot feeding upon grass, fruits, and insects ; a large grub which burrows at the foot of the acacias being said to be a favourite *bonne bouche.* Its flesh is compared to that of the Rabbit.

A specimen formerly kept in the Gardens of the Zoological Society of London, according to Waterhouse, "was very active in the evening, but usually slept during the daytime, when, sitting upon his haunches, with its head thrust between its hind legs, it appeared like a large ball of fur. It was an exceedingly savage animal, bit very severely, and would not readily unfix its hold of anything it happened to seize with its teeth. When walking, the hind legs were only used, and these were very widely separated. The tail assisted slightly in supporting the body, which was but little raised in front." The males somewhat exceed the females in size.

II. WHITE-TAILED RABBIT-BANDICOOT. PERAGALE LEUCURA.

Peragale leucura, Thomas, Ann. Mag. Nat. Hist., ser. 5, p. 397 (1887), and Cat. Marsup. Brit. Mus., p. 225 (1888).

Characters.—Probably smaller than the last ; form slender ; fur long, soft, and silky. General colour pale yellowish-fawn; under-parts and limbs white. Ears thinly clothed with fine silvery hairs; greater portion of soles of hind feet hairy. Tail moderate, slender, tapering, short-haired, and, except on the upper surface of the terminal third, uniformly white.

Distribution.—Probably Central or North-Central Australia.

This interesting species is only known from a specimen of a very young male preserved in the collection of the British Museum. While agreeing in all other essential features with the typical representative of the genus, this species makes a remarkable approximation in the structure of its upper teeth to the genus *Perameles*, and thus serves to indicate that the Rabbit-Bandicoots are in all probability a specialised offshoot therefrom.

THE TRUE BANDICOOTS. GENUS PERAMELES.

Perameles, Geoffr., Bull. Soc. Philom., vol. iii., No. 80, p. 249 (1803).

Form varying from a stout and clumsy to a light and delicate figure; muzzle long and Pig-like; ears variable in length; fore feet with the first and fifth toes short and clawless, and the three middle ones subequal, with powerful curved claws; hind feet with the first toe (hallux) short and clawless, the second and third toes with flat twisted nails, the fourth long and powerful, with a stout pointed claw, and the fifth similar but smaller. Tail tapering, short-haired, or nearly naked. Six or eight teats. Usually five pairs of upper, and three of lower incisor teeth; upper molars either triangular or quadrangular, with numerous sharp cusps.

The true Bandicoots, of which there are upwards of some

twelve species, are all comparatively small creatures, distributed over Australia, Tasmania, New Guinea, and some of the neighbouring islands. Exclusively terrestrial in their habits, and making nests of dried herbage, sticks, and leaves, they subsist chiefly on insects, grubs, worms, and bulbous and other roots, as well as fallen berries and larger fruits. In consequence of these habits they do much damage to the gardens and corn-fields of the colonists, by whom they are cordially detested. From the large quantity of earth found in their stomachs it is probable that the greater proportion of their food consists of worms.

It is mentioned by the author of the British Museum catalogue of Marsupials, that the Australian representatives of the genus fall naturally into two well-defined groups, of which the

Side View of Teeth and Jaws of a Bandicoot.

one is typified by Gunn's Bandicoot (Plate XXI.), while the other is best represented by the short-nosed species (Plate XXIII.). Had we these Australian forms alone to deal with, it might, indeed, be advisable to class these animals under two generic headings. They are, however, so closely connected by the intermediate Papuan forms that any such subdivision is impracticable; these annectent forms affording one more instance of the survival of generalised and presumably ancient types in New Guinea to which allusion has been already made.

1. STRIPED BANDICOOT. PERAMELES BOUGAINVILLII.

Perameles bougainvillii, Quoy and Gaimard, Voyage Uranie, Zool., p. 56 (1824); Thomas, Cat. Marsup. Brit. Mus., p. 246 (1888).

Perameles myosurus, Wagner, Archiv. fur. Nat., vol. vii., pt. 1, p. 293 (1841).

Perameles arenaria, Gould, Proc. Zool. Soc., 1844, p. 104.

Characters.—Size small; form light and delicate; fur coarse, but not spinose; muzzle long and slender; ears long, narrow, and pointed, reaching, when laid forward, beyond the eyes, their backs greyish flesh-colour, slightly darker on the anterior portion of the terminal half. General colour olive-grey; under-parts white; sides of rump with ill-defined dark and light transverse stripes. Soles of hind feet hairy, black posteriorly, naked, granulated, and flesh-coloured in front, with well-marked round pads at the bases of the fourth and fifth toes. Tail moderate, brown above, white beneath. Six teats. Length of head and body about 9 inches; of tail $4\frac{1}{4}$ inches.

Distribution.—West Australia.

Variety.—In South-Eastern and Southern Australia the typical form of the species is replaced by a variety (the *Perameles fasciata* of Gray) distinguished by the more marked contrast between the dark and light patches on the back of the ears, and the well-defined and conspicuous dark and light striping of the rump.

The present, together with the two following species, belongs to an exclusively Australian group, in all of which the posterior-half of the sole of the hind foot is hairy, the ear is long, and when laid forward reaches at least to the eye, and has a pointed tip, while there are eight teats.

Habits.—In common with its nearest allies, this species

frequents the stony ridges of the hotter and more exposed portions of the country, whereas the Short-nosed Bandicoot and its kindred prefer low swampy grounds covered with the densest vegetation. On the other hand, in the neighbourhood of the Swan river, Gould states that the present species "resides in the densest scrub; thickets of the seedling *Casuarinæ* being its favourite resort. It makes a compact nest in a hollow on the ground, of grasses and other materials, which assimilate so closely in colour and appearance to the surrounding herbage, that it is very difficult of detection, the difficulty being much increased by the nest having no visible opening for the ingress and egress of the animals. The nests are generally inhabited by pairs; the young are either three or four in number."

"Its food consists of insects, seeds, &c. It excavates holes in the earth, with rapidity and ease, and to these, and the hollow trunks of fallen trees, it flies for shelter when pursued by its enemies."

In regard to its burrowing habits, Krefft states that in spite of its strong claws, the Striped Bandicoot seldom digs holes except in search of its food, which comprises bulbous roots, plants, and insects. Nocturnal and social in its wild state, this animal, according to the writer quoted, bears captivity well, and becomes an adept in Mice-catching. He observes that the Bandicoot "would tumble the Mice about with its fore paws, break their hind legs, and eat generally the head only. I have seen a single individual kill as many as twenty Mice in a very short time, breaking their bones successively, after which it would satisfy its hunger."

The young are born from May to August, and are generally from two to four in number, although the female has upwards of eight teats.

In captivity a specimen lived chiefly on insects and raw meat, which were encircled by the long tongue and conveyed

to the mouth piece by piece. In consequence of the extreme delicacy and thinness of its integuments, this species is one of the most difficult of Mammals to skin.

II. GUNN'S BANDICOOT. PERAMELES GUNNI.

Perameles gunni, Gray, Proc. Zool. Soc., 1838, p. 1; Thomas, Cat. Marsup. Brit. Mus., p. 245 (1888).

(*Plate XXI.*)

Size large; form slender; fur soft, not spinose; muzzle long and slender; ears large and pointed, reaching, when laid forward, to beyond the eyes, their backs yellowish-brown, with a darker blotch on the anterior portion of the terminal half. General colour grizzled yellowish-brown; under-parts white or yellowish-white; sides of rump with four transverse vertical stripes. Soles of hind feet hairy and black posteriorly, naked and white anteriorly, with small rounded striated pads at the bases of the third and fourth toes. Tail very short and slender, white except for a short portion of the base of the upper surface. Eight teats. Length of head and body about 16 inches; of tail less than 4 inches.

Distribution.—Tasmania, and probably the coast region of South-Eastern Victoria.

This species, according to an account given by its discoverer, Ronald Gunn, is abundant in all parts of Tasmania, where it burrows in the loose soil in search of roots and worms. The harm that it inflicts on gardeners is sufficiently indicated by the circumstance that on one occasion a Bandicoot of this species completely destroyed a whole collection of imported bulbs.

The white transverse bands ornamenting its hind-quarters afford a ready means of distinguishing between Gunn's Bandicoot and the nearly allied species next mentioned.

III. LONG-NOSED BANDICOOT. PERAMELES NASUTA.

Perameles nasuta, Geoffr., Ann. Muséum, vol. iv., p. 62 (1804); Thomas, Cat. Marsup. Brit. Mus., p. 242 (1888).

Perameles lawsoni, Quoy and Gaimard, Voyage Uranie, p. 57 (1824).

(*Plate XXII.*)

Characters.—Size and form as in the preceding; fur coarse, rough, and slightly spinose; muzzle very long and slender. General colour dull olivaceous brown, without stripes on the rump; under-parts white, as are the inner sides of the limbs and the feet. Ears long, narrow, and pointed. Soles of hind feet granulated, black and thinly haired behind, white and naked in front. Tail of moderate length, brown above, paler beneath. Eight teats. Length of head and body about 16 inches; of tail 5 inches.

Distribution.—Eastern Australia.

Mr. Thomas remarks that although in its long ears, and the general structure of its skull and the form of its lower jaw, this species clearly belongs to the same group as the two preceding ones, yet in its spiny fur, the absence of stripes on the rump, and the characters of the palate and the so-called bullæ in the auditory region of the skull, it approaches so nearly to some of the Papuan species as to forbid the splitting up of the genus.

There appears to be nothing special recorded in the habits of the species, which differs from the facts already mentioned under the heading of its allies.

IV. LONG-TAILED BANDICOOT. PERAMELES LONGICAUDA.

Perameles longicauda, Peters and Doria, Ann. Mus. Genova, vol. viii., p. 335 (1876).

PLATE XXII

LONG-NOSED BANDICOOT.

Perameles longicaudata, Thomas, Cat. Marsup. Brit. Mus., p. 241 (1888).

In common with the whole of the remaining members of the genus, this species differs from the preceding group in the following points: Centre of sole of hind foot in the adult naked nearly, or quite, to the heel. Ear short, so that when laid forwards, it falls short of, or barely reaches, the eye, its tip being rounded instead of pointed.

Characters.—Size small; form light and slender; fur close, coarse, and rough, but not spiny. General colour dull greyish-brown, the thinner hairs tipped with dull yellow or grey; chin greyish-brown; under-parts, as well as inner sides of limbs, white or yellowish. Ears rounded, rather small, and broad, not reaching the eyes when laid forwards. Fore feet with brown hair, but the fingers naked; backs of hind feet white; their soles wholly naked, finely granulated, with small rudimental pads at the bases of the toes. Tail as long as the body (exclusive of the head), tapering, thinly haired, with the terminal two inches of the under surface white, elsewhere brown. Eight teats.

Distribution.—North-western New Guinea.

V. RAFFRAY'S BANDICOOT. PERAMELES RAFFRAYANA.

Perameles raffrayana, Milne-Edwards, Ann. Sci. Nat., ser. 6, vol. vii., pl. viii. (1878); Thomas, Cat. Marsup. Brit. Mus.; p. 239 (1888).

Characters.—Size rather large; form light and delicate; head long and slender; fur thick, close, rough, and slightly spinous in places. General colour coarsely grizzled rufous or umber-brown, with the thicker hairs black, and the thinner ones tipped with orange or rufous; head dark brown, with the sides of the face paler; chin and chest white; rest of under-parts white or white and brown; limbs brown, with the fingers nearly naked. Muzzle very long and slender, with the top and sides of the nose naked. Ears longer and narrower than in other members of the group, not quite reaching the eyes when turned forwards, and with rounded tips. Soles of hind feet naked and coarsely granulated, with indistinct rudimental pads at the bases of the fourth and fifth toes. Tail long, cylindrical, and uniformly clothed with short fine black hairs. Eight teats.

Distribution.—New Guinea.

VI. BROADBENT'S BANDICOOT. PERAMELES BROADBENTI.

Perameles broadbenti, Ramsay, Proc. Linn. Soc. N. South Wales, vol. iii., p. 402 (1878); Thomas, Cat. Marsup. Brit. Mus., p. 240 (1888).

Founded upon a very large adult male specimen which is only provisionally admitted by the author of the British Museum Catalogue of Marsupials as specifically distinct from the preceding; its chief claim to distinction being based on the peculiar character of the tail, which is stated to differ from that of all other members of the family in being partially prehensile. This appendage is long, scaly above, and covered below with flattish transverse scaly tubercles; its colour being blackish for about two-thirds of its length, and thence fawn-coloured to the tip, with a sparse clothing of short hair. The great size of the type specimen may be merely due to sex and age.

Distribution.—South-eastern New Guinea.

VII. DORIA'S BANDICOOT. PERAMELES DOREYANA.

Perameles doreyana, Quoy and Gaimard, Voyage Astrolabe, Zool., vol. i., p. 100 (1830); Thomas, Cat. Marsup. Brit. Mus., p. 236 (1888).

THE TRUE BANDICOOTS. 143

Perameles rufescens, Peters and Doria, Ann. Mus. Genov., vol. vii., p. 541 (1875).

Perameles aruensis, Peters and Doria, *op. cit.*, p. 542.

Characters.—This and the following species may be at once distinguished from all the other members of the genus by having only four, in place of five, pairs of upper incisor teeth. The present species is characterised by its large size and elongated muzzle, coupled with the presence of eight teats; its general colour being dark coppery-brown, coarsely grizzled with orange.

Distribution.—New Guinea and neighbouring islands, including the Kei and Aru groups.

VIII. COCKERELL'S BANDICOOT. PERAMELES COCKERELLI.

Perameles cockerelli, Ramsay, Proc. Linn. Soc. N. South Wales, vol. i., p. 310 (1877); Thomas, Cat. Marsup. Brit. Mus., p. 238 (1888).

Perameles myoides, Günther, Ann. Mag. Nat. Hist., ser. 5, vol. xi., p. 247 (1883).

Perameles garagassi, Mikl. Maclay, Proc. Linn. Soc. N. South Wales, vol. ix., p. 715 (1884).

Distinguished from the last by its inferior size, and shorter muzzle, coupled with the presence of only six mammæ; the general colour being coarsely grizzled black and yellow.

Distribution.—North coast of New Guinea, Aru, and adjacent islands.

IX. NORTH AUSTRALIAN BANDICOOT. PERAMELES MACRURA.

Perameles macrura, Gould, Proc. Zool. Soc., 1842, p. 41; Thomas, Cat. Marsup. Brit. Mus., p. 234 (1888).

The skulls of this and the remaining species of the genus

differ from those of all the preceding forms by the large size and pear-shaped contour of the auditory or tympanic bullæ, situated below the aperture of the internal ear. In the whole of the other species these bullæ are small, hemispherical, and often more or less imperfectly ossified.

Characters.—Size large; form rather stout; fur short, coarse, and spiny. General colour coarsely grizzled yellow and black; under-parts white or yellowish-white. Ears short and broad, their backs brown, narrowly margined with white. Feet brown, or mixed brown and white; soles of hind feet naked and coarsely wrinkled. Tail rather long, brown above and white beneath. Eight teats. Length of head and body about 16 inches; of tail 7 inches.

Distribution.—Northern Australia.

X. GOLDEN BANDICOOT. PERAMELES AURATA.

Perameles auratus, Ramsay, Proc. Linn. Soc. N. South Wales, ser. 2, vol. ii., p. 551 (1887).

Perameles aurata, Ogilby, Cat. Austral. Mamm., p. 23 (1892).

Characters.—Size small; form rather stout; fur coarse and spiny. General colour rich golden-brown pencilled with black; under parts white. Length of head and body about 8½ inches.

Distribution.—North-Western Australia.

In the British Museum Catalogue of Marsupials this Bandicoot was identified with the preceding, from which (assuming the type specimen to be adult) it appears sufficiently distinguished by its greatly inferior size.

XI. SHORT-NOSED BANDICOOT. PERAMELES OBESULA.

Didelphis obesula, Shaw, Nat. Miscell., vol. viii., pl. ccxcviii. (1793).

PLATE XXIII

SHORT-NOSED BANDICOOT

Perameles obesula, Geoffroy, Ann. Muséum, vol. iv., p. 64 (1804); Thomas, Cat. Marsup. Brit. Mus, p. 231 (1888).

Perameles fusciventer, Gray, in Grey's Australia, Appendix, vol. ii , p. 407 (1841).

Perameles affinis, Gray, List Mamm. Brit. Mus., p. 96 (1843).

(*Plate XXIII.*)

Characters.—Size medium ; form stout. In all other respects externally similar to *Perameles macrura*, except that the tail is shorter, the feet are rather less heavy, and the colour is lighter. Length of head and body about 14 inches ; of tail 5½ inches.

Distribution.—Australia, south of the tropics, and Tasmania.

Habits.—It does not appear that anything special has been recorded of the habits of this and the allied species whereby they can be distinguished from Gunn's Bandicoot and its kindred, with the exception that, as already mentioned, they are stated by Gould to inhabit damper and more densely wooded situations.

XII. PORT MORESBY BANDICOOT. PERAMELES MORESBYENSIS.

Perameles moresbyensis, Ramsay, Proc. Linn. Soc. N. South Wales, vol. ii., p. 14 (1878).

Characters.—General external characters as in *P. obesula* and *P. macrura*, except that the fur is coarser and more spiny, the colour of the crown of the head darker, and the hinder portion of the back dark grizzled orange, instead of being uniform with the fore-quarters.

Distribution.—South-eastern New Guinea.

Although the above-mentioned points of distinction from the Australian forms are small they appear, according to Mr. Thomas, to be constant, and as the Papuan form inhabits an area separated from that of the others by a wide stretch of sea, it may be provisionally allowed to rank as a species.

THE PIG-FOOTED BANDICOOTS. GENUS CHŒROPUS.

Chœropus, Ogilby, Proc. Zool. Soc., 1838, p. 25.

Form light and slender; muzzle short and narrow; ears long and narrow; fore feet with the first and fifth digits wanting, the fourth rudimentary, and the second and third fully developed, with long, slender claws; hind feet with the first toe (hallux) wanting, the fifth rudimentary, and the fourth very large; tail cylindrical, slightly crested along the upper surface. Eight teats. Five pairs of upper, and three of lower incisor teeth.

The single Australian representative of this remarkable genus is approached, observes Mr. Thomas, both in general external appearance, and in the structure of the skull, by Gunn's Bandicoot and its allies, from which group it may therefore probably be regarded as a specialised offshoot. "Its distinction as a genus is, however, unquestionable, even if it were not for the unique and peculiar structure of its fore feet, which have such a striking resemblance to those of the Pig as to have gained for the only species its common English name of Pig-footed Bandicoot."

1. PIG-FOOTED BANDICOOT. CHŒROPUS CASTANOTIS.

Chœropus ecaudatus, Ogilby, Proc. Zool. Soc., 1838, p. 25.
Chœropus castanotis, Gray, Ann. Mag. Nat. Hist., vol. ix., p. 42 (1842); Thomas, Cat. Marsup. Brit. Mus., p. 250 (1888).
Chœropus occidentalis, Gould, Mamm. Australia, vol. i., pl. vi. (1845).

Characters.—Size small; form delicate; fur coarse and straight, but not spiny. Ears thinly haired, dull chestnut-brown behind, darkening towards the tip. General colour coarse grizzled grey, with a tinge of fawn; under-parts white. Limbs long and slender; tail short, black above, grey on the sides and beneath. Length of head and body about 10 inches; of tail 4 inches.

Distribution.—Western New South Wales and Victoria, and South and West Australia.

Habits.—First described from an accidentally mutilated example, this curious animal was supposed to be tailless, and was, therefore, named *Chæropus ecaudatus*, but on the acquisition of other specimens, showing the presence of a well-developed tail, the name was very properly changed to *castanotis*, in allusion to the chestnut colour of the ears. In size the Pig-footed Bandicoot may be compared to a small Rabbit. Writing of its first discovery in 1836, Mitchell observes that "the most remarkable incident of this day's journey was the discovery of an animal of which I had seen only a head in a fossil state in the limestone caves of the Wellington Valley, where, from its very singular form, I supposed it to belong to some extinct species. The chief peculiarity then observed was the broad head and very long slender snout, which resembled the narrow neck of a wide bottle; but in the living animal the absence of a tail was still more remarkable. The feet, and especially the fore legs, were also singularly formed, the latter resembling those of a Pig; and the Marsupial opening was downwards and not upwards, as in the Kangaroo and others of that class of animals. This quadruped was discovered by the natives on the ground; but on being chased it took refuge in a hollow tree, from which they took it alive, all of them declaring that they had never before seen an animal of the kind. This was when the party had commenced the journey up the left bank of the Murray, immediately after crossing that river."

The most remarkable statement in this account is the alleged occurrence of fossilised remains of this animal in the caves of the Wellington Valley, since, among all the vast numbers of bones subsequently acquired from that district by the British Museum, none are referable to the Pig-footed Bandicoot. It may be remarked, in passing, that from the structure of this family, it might have been thought that the backward direction of the opening of the pouch was the original condition among the Marsupials, and that its forward direction in the Kangaroos, and other forms which habitually maintain a more or less nearly upright position, had been acquired in order to prevent the helpless young from falling out. The condition prevailing in the *Dasyuridæ* disproves, however, this view. Burrowing in the ground like the other members of the family, the Pig-footed Bandicoot seems to be omnivorous in its tastes, but appears to be specially fond of insects. Sturt states that in the Darling district these animals are generally found lying out in the grass, and that when chased by dogs they almost invariably took refuge, after a short run, in hollow logs, from which they were readily cut out. In the open they squat like Rabbits, laying back their broad ears along the shoulders in a very similar manner. When confined in a box, captured specimens ate sparingly of grass and tender leaves, although showing a much greater partiality for flesh. The latter diet did not, however, appear to suit their constitution, all the specimens dying in succession.

Although still found thirty years ago on the plains of the Murray district, Krefft states that it was even then rapidly becoming rare, owing to the increase of cattle and sheep. After much trouble, that observer succeeded in securing some living specimens, whose history he recounts as follows: "About sundown, when I was about to secure my animals for the night, one of the nimblest made its escape, jumping clean through the wires of its cage. At a quick pace it ran up one of the sandstone cliffs, followed by myself, all the black-fellows, men, women, and children, and their dogs. Here was a splendid opportunity for observing the motions of the animal, and I availed myself of it. The *Chœropus* progressed like a broken-down hack in a canter, apparently dragging the hind-quarters after it; we kept in sight of the fugitive, and, after a splendid run up and down the sand-hills, our pointer, who had been let loose, brought it to bay in a salt-bush. A large tin-case was fitted up for the habitation of these animals, and provided with coarse barley-grass, upon which, as the natives informed me, they feed. Insects, particularly grasshoppers, were also put into the box, and, though they were rather restless at first, and made vain attempts to jump out, they appeared snug enough in the morning, having constructed a completely covered nest with the grass and some dried leaves.

"During the daytime, they always kept in their hiding-places, and, when disturbed, quickly returned to them; but, as soon as the sun was down, they became lively, jumping about and scratching the bottom of the case in their attempts to regain liberty. I kept these animals upon lettuces, barley-grass, bread, and some bulbous roots, for six weeks."

The Pig-footed Bandicoot is a free drinker, but never attacks Mice, after the manner of the other members of the family. Although provided with eight teats, the female produces only a pair of young at a birth, these probably making their appearance in May or June. It is noteworthy that in the young animal the third toe of the fore foot is relatively more developed than in the adult. It does not appear that this animal has ever been exhibited alive in England.

FAMILY DASYURIDÆ. THE DASYURES, THYLACINES, ETC.

Limbs subequal; fore feet with five toes; hind feet with the third and fourth toes completely separate, the first (hallux) small and clawless, or wanting, and the others of subequal size. Tail long, hairy, and non-prehensile. Stomach simple; intestine without a blind appendage or cæcum; pouch, when present, opening forwards or downwards. Four pairs of upper, and five of lower incisor teeth; and the whole of the dentition of a thoroughly carnivorous type, the upper molars being more or less triangular in form, and carrying a number of sharp cusps.

This family, which is distributed over Australia and Tasmania, New Guinea, and the adjacent islands, exclusive of those of the Austro-Malayan region, embraces the only thoroughly carnivorous Australian Marsupials, and likewise the largest representatives of the Polyprotodont subdivision of the order. Although all its members subsist on an animal diet, the smaller kinds are either wholly or partially insectivorous. None appear to be carnivorous.

As a whole, the *Dasyuridæ* may be regarded as among the most generalised of all living Marsupials, some of them, and more especially the little Banded Anteater, retaining indications of affinity with the extinct Jurassic Marsupials of Europe which are unknown elsewhere.

THE THYLACINES GENUS THYLACINUS.

Thylacinus, Temminck, Monogr. Mamm., vol. i., p. 60 (1827).

This genus, together with all the other members of the family, with the exception of the Banded Anteater, constitute a sub-family (*Dasyurinæ*), characterised as follows: Tongue short, simple, and non-extensile; lower lip rounded and not produced; chest without a gland. Four pairs of large molar teeth in each jaw, the lower ones with their outer cusps larger than the inner ones.

The following may be taken as the distinctive characteristics of the genus under consideration: Size large; form Wolf-like; muzzle long and slender; ears of medium size; tail long, short-haired; feet markedly digitigrade, the front ones with the toes furnished with short, thick, conical claws, and the hind pair with only four toes, owing to the absence of the hallux. Back marked with transverse black bands. Four teats. Marsupial bones rudimentary. Four pairs of premolar teeth.

Although the occurrence in Queensland of an animal allied to the Thylacine has been reported, this has not been confirmed by the capture of specimens, and the genus seems now to be represented only by the undermentioned Tasmanian species. There is, however, good evidence that a larger species existed during the Pleistocene, or latest geological epoch, on the Australian mainland, where its fossilised remains are not of uncommon occurrence in the caverns and superficial deposits.

I. TASMANIAN THYLACINE. THYLACINUS CYNOCEPHALUS.

Didelphys cynocephala, Harris, Trans. Linn. Soc., vol. ix., p. 174 (1808).
Dasyurus cynocephalus, Geoffroy, Ann. Muséum, vol. xv., p. 304 (1810).
Thylacinus harrisi, Temminck, Monogr. Mamm., vol. i., p. 63 (1827).
Thylacinus cynocephalus, Fischer, Syn. Mamm., p. 270 (1829); Thomas, Cat. Marsup. Brit. Mus., p. 255 (1888).

Thylacinus breviceps, Krefft, Ann. Mag. Nat. Hist., ser. 4, vol. ii., p. 296 (1868).

(*Plate XXIV.*)

Characters.—Fur short, close, and crisp. General colour pale, finely grizzled, greyish-brown, with a faint yellowish or tawny tinge; under-parts somewhat paler; edges and bases of the ears, as well as a patch round each eye, nearly white; hinder part of back marked with some sixteen blackish-brown transverse bands, descending in the region of the rump nearly to the knee. Soles of hind feet naked and coarsely granulated, without distinct pads. Tail with indistinct crests of hair on its upper and lower surfaces, with its tip blackish. Length of head and body about 44 inches; of tail 21 inches.

It may be mentioned as a somewhat remarkable fact that traces of a rudimental pouch are found in the male Thylacine, and it should be added that such a rudimental pouch, which may be either permanent or transitory, has been detected in the males of several other Marsupials, most of which belong to the Polyprotodont division of the order.

Distribution.—Tasmania.

Habits.—Known among the colonists by the names of Native Wolf, Tiger, or Hyæna, the Thylacine was at one time an abundant animal in its native island. The damage which it inflicts on the flocks of the settlers has, however, given rise to a relentless war of extermination, which has resulted in the almost complete extinction of this, the largest of the Australasian Carnivores, in the more settled portions of the country.

So like in general appearance to a Wolf is this animal, that the name of Tasmanian Wolf might well receive general adoption, were it not for the circumstances that the application of the name of a placental Mammal to a Marsupial is best, when possible, avoided. And on this ground alone we prefer to

PLATE XXIV

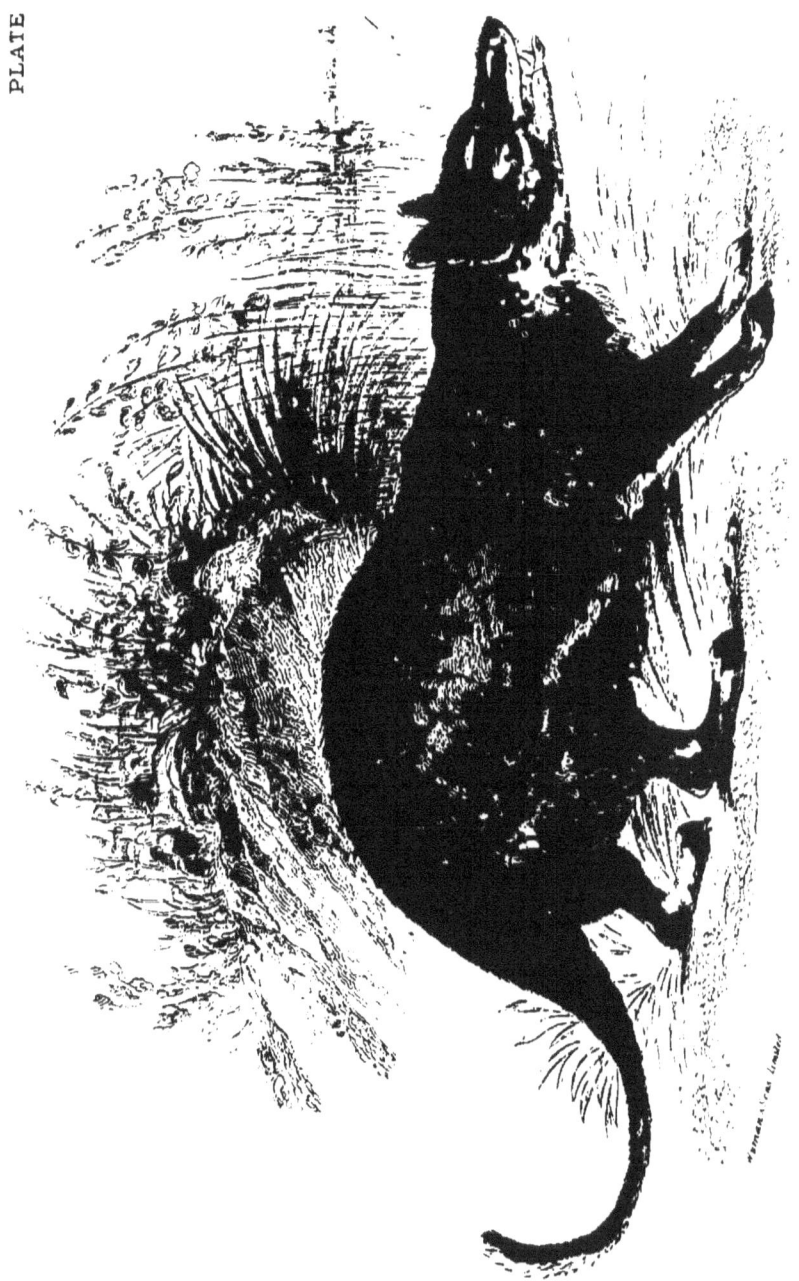

use an Anglicised form of its scientific designation, as the popular name of this animal.

The Thylacine appears to be generally found among caverns and rocks in the deep and almost impenetrable glens in the neighbourhood of the highest mountains of Tasmania. Chiefly nocturnal in their habits, these animals are dull and inactive on the rare occasions when they show themselves by daylight, moving with a slow pace, and incessantly blinking from the unaccustomed light. Their cry appears to be limited to a dull guttural growl; and it seems that, unlike Wolves, they never hunt in packs. Before the introduction of flocks into the country, the Thylacine doubtless subsisted mainly on the smaller Kangaroos and Wallabies, together with other Marsupials; the first known specimen, captured by Harris, having portions of a spiny Anteater in its stomach. Sheep are, however, easier animals to kill than Kangaroos, and consequently in the more settled parts of the country the Thylacine soon took to sheep-killing. Its depredations on the flocks are always effected during the night-time; and some idea of its ferocity may be obtained from a statement of Gunn to the effect that in the case of large old males even several dogs together will refuse to make an attack.

The female Thylacine usually produces four young ones at a birth; the presence of only four teats in the pouch of the female indicating that this number must be the limit.

THE URSINE DASYURES. GENUS SARCOPHILUS.

Sarcophilus, F. Cuvier, Hist. Nat. Mamm., vol. iv., pt. 70 (1837).

Form very stout and powerful; muzzle short and broad; ears broad and rounded; tail moderate, evenly haired; feet plantigrade, the front ones with well-developed curved claws, and in the hinder pair the hallux wanting. Soles of hind feet naked, without well-defined pads. Body blotched with white. Two pairs of premolar teeth; the upper molars, save the last, very strong, triangular in form, and much shorter and wider than in the preceding genus.

Like the latter, the genus *Sarcophilus* is now represented only by a single Tasmanian species, although a second one, which became extinct before the historic period, formerly inhabited the mainland of Australia. The affinities of the genus appear to be closer with *Thylacinus* than with *Dasyurus*, although by many writers the Tasmanian Devil is placed among the latter animals.

I. THE TASMANIAN DEVIL. SARCOPHILUS URSINUS.

Didelphis ursina, Harris, Trans. Linn. Soc., vol. ix., p. 176 (1808).

Dasyurus ursinus, Geoffroy, Ann. Muséum, vol. xv., p. 305 (1810).

Sarcophilus ursinus, F. Cuvier, Hist. Nat. Mamm., vol. iv., pt. 70 (1837); Thomas, Cat. Marsup. Brit. Mus., p. 259 (1888).

Characters.—Fur thick and close, consisting largely of soft woolly under-fur. General colour black or blackish-brown, with a variable number of white patches on the neck, shoulders, rump, and chest, the latter being alone constant. Ears hairy, with well-marked tufts at the base; soles of feet marked, coarsely granulated, and without pads, but a small transversely striated pad at the tip of each toe; tail short, and uniformly thickly haired. Length of head and body about 28 inches; of tail 12 inches.

Distribution.—Tasmania.

Habits.—Like the Thylacine, the Tasmanian Devil, as the animal under consideration, from its ferocity and destructive habits, is universally termed by the colonists of its native island, was first made known to science by Harris in the year 1808. In his original account he writes that "these animals were very common on our first settling at Hobart Town, and were particularly destructive to poultry, &c. They, however, furnished the convicts with a fresh meal, and the taste was said to be not unlike veal. As the settlement increased, and the ground became cleared, they were driven from their haunts near the town to the deeper recesses of the forests yet unexplored. They are, moreover, easily secured by setting a trap in the most unfrequented parts of the woods, baited with raw flesh, all kinds of which they eat indiscriminately and voraciously;

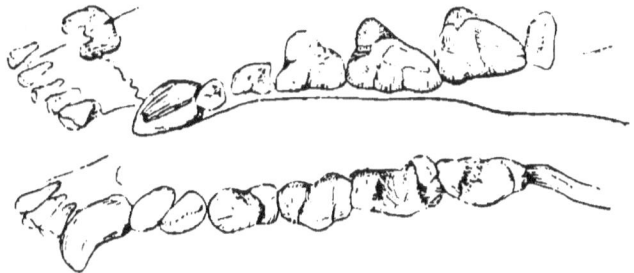

Right Upper and Lower Teeth of the Tasmanian Devil.

they also, it is probable, prey on dead fish, blubber, &c., as these are frequently found on the sand of the sea-shore.

"In a state of confinement they appear to be untameably savage, biting severely, and uttering at the same time a low yelling growl. A male and female, which I kept for a couple of months chained together in an empty cask, were continually fighting. Their quarrels began as soon as it was dark (as they slept all day) and continued throughout the night almost without intermission, accompanied by a kind of hollow barking, not unlike that of a dog, and sometimes a sudden kind of snorting, as if the breath was retained a considerable time and then suddenly expelled. They frequently sat on their hind parts and used their fore paws to convey food to their mouths. The muscles of the jaws were very strong, as they cracked the largest bones with ease asunder."

Exclusively carnivorous in their habits, these animals, despite their comparatively small size, commit great havoc among the sheepfolds of the settlers, and are stated to be so ferocious and to bite with such severity that one of them is fully a match for any ordinary Dog. It is a curious comment on the present state of zoological knowledge that naturalists are still unacquainted with the number of teats in the Tasmanian Devil. Since, however, the female is stated to produce from three to five young ones at a birth it is probable that there are six teats. Beyond this statement as to the number of the young, nothing seems to have been ascertained with regard to the breeding habits of these creatures. Like many of its kindred, the Tasmanian Devil is a burrowing and nocturnal animal. In size it may be compared to a Badger, and owing to its short limbs, plantigrade feet, and short muzzle, its gait and general appearance are very Badger- or Bear-like.

In order to afford some further notion of the ferocity of these animals, we may quote the following passage from Krefft: "One of them, and by no means a large one, escaped not long ago, and killed in two nights fifty-four Fowls, six Geese, an Albatross and a Cat. Having been recaptured in what was considered a stout trap, with a door constructed of iron bars as thick as a lead-pencil, he made his escape by twisting this solid obstacle aside, almost doubling it up with his powerful teeth. To give some idea of the strength of the animal, we mention that the blacksmith who repaired the trap could not bend the bars back into their position without proper tools. When caught in a Fox-trap, the black Dasyure often bites off the fastened limb and escapes. A specimen in the Museum collection, the largest ever secured, had only three legs, one of the hind limbs was clean gone, and not even the trace of a stump remained visible."

The enormous crushing power of the teeth of this animal has already been alluded to, and will be fully borne out by an inspection of the skull. The cheek-teeth, as shown in the figure on page 155, are especially characterised by their stoutness and close crowding together, the latter feature being due to the shortness of the muzzle. The middle pair of upper incisor teeth are larger than those next them, while the tusks or canine teeth are very large and powerful.

The skeletons of both the Thylacine and the Tasmanian Devil may be distinguished from those of all the other members of the family by the presence of a perforation on the inner side of the lower end of the humerus, or upper bone of the fore limb. In a skeleton of the present species in the Museum of the Royal College of Surgeons it is, however, stated that there is a perforation in one humerus and not in the other. The absence of the same perforation in the Dasyures is a very unusual feature among Marsupials, nearly all of which retain indications of their affinity with certain extinct reptiles by its presence.

THE DASYURES. GENUS DASYURUS.

Dasyurus, Geoffroy, Bull. Soc. Philom., vol. i., p. 106 (1796).

General form usually slender and Weasel- or Civet-like, but occasionally stouter. Ears long and narrow; nose naked, with a deep groove running down to the upper lip; tail long, uniformly and thickly haired throughout; feet plantigrade, with sharp, curved claws; the first toe, or hallux, of the hind foot very small or wanting; soles of the feet granulated, and nearly or wholly naked. Body spotted. Pouch opening vertically downwards; six or eight teats. Two pairs of premolar teeth in each jaw; and all the teeth much weaker than in the preceding genus.

To the settlers the members of this genus are commonly known by the name of Native Cats, but since such names, as we have already had occasion to mention, are decidedly objectionable, we prefer to take an Anglicised form of the scientific title. These animals range not only over Australia and Tasmania, but likewise occur in New Guinea and the adjacent islands. Unlike the Thylacine and Tasmanian Devil, which are purely terrestrial, the majority of the Dasyures are more or less arboreal in their habits; while they are both carnivorous and insectivorous. Mr. Thomas suggests, however, that certain species (*Dasyurus viverrinus* and *D. geoffroyi*), in which distinct and striated pads are lacking on the soles of the feet, are probably far less arboreal than the others, since the organs in question seem to be developed *pari passu* with the scansorial powers of their possessors. Obnoxious, and at the same time well known, to the settlers on account of their depredations to the hen-roost and the dove-cot, the Dasyures may be regarded as playing in Australia the *rôle* of the Martens and Weasels in Europe, since they subsist very largely upon birds and, at one season of the year, on their eggs.

1. SPOTTED-TAILED DASYURE. DASYURUS MACULATUS.

Viverra maculata, Kerr, Linn. Anim. Kingdom, p. 170 (1792).
Dasyurus macrourus, Geoff., Ann. Muséum, vol. iii., p. 358 (1804).
Dasyurus maculatus, Fischer, Zoogn., vol. ii., p. 584 (1813); Thomas, Cat. Marsup. Brit. Mus., p. 263 (1888).

(*Plate XXV.*)

Characters.—Size large; form stout and heavy; fur thick and close. General colour dark brown (never black) with a rufous

SPOTTED-TAILED DASYURE.

or orange tinge, and with large white spots; under-parts white or pale yellow. Ears rather short or very thinly haired. Hind feet with the first toe, or hallux, present; claws of both feet large and powerful. Tail very long, brown or rufous brown, and spotted like the body. Six teats. Length of head and body about 25 inches; of tail 19 inches.

Externally this species may be distinguished from all the other members of the genus by its large size and spotted tail; while it is further characterised by the presence of well-defined and transversely striated pads on the soles of the feet. Its skull differs from that of every other species by the auditory or tympanic bullæ at the base of the aperture of the internal ear being obliquely oval, instead of spherical.

Distribution.—From Central Queensland to Victoria, principally on the mountain ranges, but extending to the coast and Tasmania. In commenting on its distribution, Mr. Thomas observes that "the commonness of this species in Tasmania and its great rarity on the Continent are of interest as showing that it is appproaching the condition now exhibited by the Thylacine and Tasmanian Devil, namely, complete extermination in Australia, where both once lived, and continued preservation in the island of Tasmania. Why the small island of Tasmania should be able to support in considerable numbers the three largest carnivorous Marsupials, competing probably, to a certain extent, with each other, while they have been almost or wholly unable to survive on the mainland, is a problem which much requires elucidation. The presence of the Dingo in Australia is no doubt one of the causes that have tended to produce this remarkable state of affairs."

To this statement Mr. Ogilby replies by traversing the assertion that the species under consideration is rare on the Australian mainland. "As a matter of fact," he writes, "*D. maculatus* is by no means uncommon—nor seemingly has it any present intention of dying out—in the mountainous and coastal districts of Eastern Australia from Northern Queensland, through New South Wales and Victoria to South, and possibly West Australia. It may be worth mentioning that the largest, stoutest, and heaviest example I have yet seen was caught, in company with five others, on Manby Beach, a suburb of Sydney. For these and other reasons I cannot in any wise agree with Mr. Thomas as to the approaching extermination of this species on the mainland, nor can I allow, though confessedly unable to promulgate a more ostensible theory, that the causes which adduced to the annihilation, at what must have been a very recent period, of *Sarcophilus* and *Thylacinus* from Eastern Australia, can have in any degree affected *D. maculatus*, the former having been purely, or at least, mainly terrestrial, while the latter is most emphatically an arboreal Mammal. If the Dingo, as suggested by Mr. Thomas, had anything whatever to do with the extermination of our Native Cats, the first to disappear would have been *D. viverrinus*, by far the most terrestrial of all the Dasyures."

In size this species may be roughly compared to a Domestic Cat; and its general habits are doubtless similar to those of the other arboreal members of the genus. Its skin is but little valued by furriers.

II. SLENDER DASYURE. DASYURUS GRACILIS.

Dasyurus gracilis, Ramsay, Proc. Linn. Soc. N. South Wales, ser. 2, vol. iii., p. 1296 (1888); Ogilby, Cat. Austral. Mamm., p. 17 (1892).

Characters.—Size small; form light and graceful; fur short, close, and somewhat harsh to the touch. General colour deep blackish-brown, spotted with white, the spots on the sides and on the basal third of the tail being larger than elsewhere, and sometimes running into one another. Ears short, thinly haired externally at the base, internally with a tuft of long hairs at the front edge. Hind feet with the first toe, or hallux ; claws long and powerful. Tail long, slender, tufted at the extreme tip, and coloured like the body. Length of head and body 13 inches ; of tail $9\frac{1}{2}$ inches.

Distribution.—Bellenden Ker Range, Northern Queensland.

Of this reputed species, which was described after the British Museum Catalogue of Marsupials was written, Mr. Ogilby remarks that " were it not for the indisputably adult dentition of the unique specimen on which Dr. Ramsay has founded his new species, and that evidence, presumably reliable, points to the existence in the same district of a Spotted-tailed Dasyure as large as or even larger than the southern *D. maculatus*, I should have been inclined to treat this specimen as merely an aborted tropical form of that species. Until, however, further research has undeniably proved the presence there of two so widely separated races it is perhaps better to keep them apart. It is worth mentioning that both in its fauna and flora the Bellenden Ker Ranges show more distinct affinities to the Papuan than to the restricted Australian sub-region. For instance, the rhododendron flourishes in a wild state in these mountains of Australia only, having evidently travelled round from the Himalayas along the highlands of New Guinea, and so down the Northern Queensland ranges: similarly such typically Papuan forms as *Dendrolagus* among Mammals, *Casuarius* among Birds, *Papuina* among Molluscs, *Perichæta* among Earthworms, with many others, have found their way into the Australian fauna."

III. COMMON DASYURE. DASYURUS VIVERRINUS.

Didelphis viverrinus, Shaw, Gen. Zool., vol. i., pt. 2, p. 491 (1800).

Dasyurus viverrinus, Geoffr., Ann. Muséum, vol. iii., p. 360 (1804); Thomas, Cat. Marsup. Brit. Mus., p. 265 (1888).
Dasyurus maugei, Geoffr., *op. cit.*, p. 359.
Dasyurus guttatus, Desmarest, Nouv. Dict. d'Hist. Nat., vol. xxiv., p. 10 (1804).

(*Plate XXVI.*)

Characters.—Size medium; form slender; fur thick and soft. General colour either pale grey or black, spotted with white. Ears large. Hind foot without the first toe (hallux); soles of feet granulated, without distinct pads. Tail bushy, its basal three-fourths coloured like the back, but without spots, the tip white. Length of head and body about 18 inches; of tail 12 inches.

The most interesting peculiarity of this species is the frequent occurrence of (melanistic) specimens in which the ground-colour of the fur is black, instead of the normal grey. Although the white spots are as fully developed in the black as in the grey variety, the tip of the tail in the former becomes of the same sooty hue as the body. For a long time the two were regarded as distinct species, but according to Gould both the grey and black forms have been found in the same litter, although, as we shall see below, this must be an unusual circumstance, as it is contrary to the experience of another observer.

Distribution.—The eastern watershed of New South Wales. Victoria, South Australia, and Tasmania.

Habits.—Writing of the habits of the Common Dasyure, Mr. Ogilby observes that "this species, in both varieties, is as much, if not more terrestrial than arboreal, living in dead rocks, or in holes in the cliffs, in which latter place they feed on dead fish, and probably crustaceans, molluscs, &c., and are thus fre-

COMMON DASYURE

quently caught in baited fish-traps left bare by the tide, or hauled up during bad weather."

In writing of this species, Krefft observes that "these little creatures, with their fierce disposition, are familiar to the greater number of colonists. They inhabit our forests; but prefer to take up their abode with civilised man when they find out that he keeps plenty of meat about his habitation, or rears poultry. They are very savage for their size, and five of them kept in a cage without sustenance for a day only, had almost reduced themselves to the state of the famous tabbies of Kilkenny. The fact is, they devoured each other till only a part remained, and the savage look and watchfulness of these two animals was amazing to behold.

"They are stubborn in the extreme, and appear to care about nothing. We have noticed them to come quite unconcerned into a tent at night and take up a cosy place near the chimney, from which a fire-stick only could dislodge them. Another case was mentioned not many days ago, when one of the Tiger-Cats actually faced a half-caste man, who was terror-stricken, and ran away. A real aboriginal native, one of the old tribes, would have made short work of such an adversary; but these poor people have now almost died out, and the few still lingering behind cannot even remember the animals which their ancestors hunted."

Like its kindred, this Dasyure is very destructive to birds, especially poultry, while it also consumes large numbers of eggs. Living during the day in hollow logs, log-fences, or holes in the ground, these marauders sally forth at night to seek their food on the ground. Here they are frequently put up by the dogs of the sportsmen in pursuit of Phalangers, and when chased they nearly always take refuge in the smaller trees, but rarely ascending the gum-trees. The belts of timber on the edges of the swamps are among their favourite haunts; while, as already mentioned, they may not unfrequently be met with on the shore. As many as six young ones may frequently be met with in a nest; and it appears that, after leaving the teats, these are carried but a very short time in the pouch.

An anonymous writer observes that in some parts of the country the black variety is very rare and local; and that he never found black and grey young ones in the same nest.

In size this species may be roughly compared to the common European Marten. The fur being soft, the skins are suitable for linings; and from two to five thousand skins are annually imported into England. Formerly the grey skins fetched from about fivepence to sixpence each in the market, while the value of the black ones ranged from tenpence to a shilling. Of late years, however, there has been a fall in the price.

IV. BLACK-TAILED DASYURE. DASYURUS GEOFFROYI.

Dasyurus geoffroyi, Gould, Proc. Zool. Soc., 1840, p. 151; Thomas, Cat. Marsup. Brit. Mus., p. 268 (1888).

Characters.—Size medium; form slender; fur thick and soft. General colour olive-grey, tinged with rufous and spotted with white; under-parts white. Hind foot with a first toe (hallux); soles of feet granulated, and their pads marked by rounded unstriated prominences. Ears large, brown at the back, with white edges. Tail long and rather bushy, the basal half above and the basal fourth below coloured like the back, but devoid of spots, the remainder black. Length of head and body about 16 inches; of tail 12 inches. Six teats.

Distribution.—All Australia, with the exception of the extreme north, and the coast districts of the south-east.

The two sexes differ somewhat in size, and it appears that the race inhabiting the western portion of the continent attains larger dimensions than the one from the opposite side. From the structure of the pads on its feet, this species is probably largely terrestrial in its habits.

V. NORTH AUSTRALIAN DASYURE. DASYURUS HALLUCATUS.

Dasyurus hallucatus, Gould, Proc. Zool. Soc., 1842, p. 41; Thomas, Cat. Marsup. Brit. Mus., p. 269 (1888).

Characters.—Size small, not more than half the bulk of *D. viverrinus*; form slender; fur short and coarse. General colour yellowish-brown, spotted with white; under-parts pale grey or yellow. Ears large, thinly clothed with fine yellow hairs. Hind foot with the first toe (hallux); soles of feet with smooth, well-defined, and transversely striated pads. Tail long, rather short-haired, at the base coloured like the body, but unspotted, elsewhere black. Eight teats. Length of head and body about 11 inches; of tail 8 inches.

Distribution.—Tropical Australia.

This species, which derives its name from the presence of the first toe or hallux in the hind foot, differs from the two preceding ones, and thereby resembles the next, in its distinct transversely striated foot-pads, which may be taken as an indication of its arboreal habits. In its small size, slender build, and the structure of the hind feet, it makes a certain approach to the members of the next genus, although in other respects it agrees with the typical Dasyures. The near relationship of this species to the next affords another instance of the affinity between the fauna of North Australia and New Guinea.

VI. PAPUAN DASYURE. DASYURUS ALBOPUNCTATUS.

Dasyurus albopunctatus, Schlegel, Notes Leyden, Mus., vol. ii., p. 51 (1880); Thomas, Cat. Marsup. Brit. Mus., p. 271 (1888).

Dasyurus fuscus, Milne-Edwards, Comptes Rendus., vol. xc., p. 1518 (1880).

Characters.—General size, structure, and coloration as in the last, from which this species may be distinguished by its rather stouter build, shorter muzzle and ears, shorter and more woolly fur, and the rufous or fulvous tinge in the general colour of the back.

Distribution.—North-western New Guinea.

THE POUCHED MICE. GENUS PHASCOLOGALE.

Phascologale, Temminck, Monogr. Mamm., vol. i., p. 56 (1827).

Body unspotted; form slender and graceful; ears rounded; tail long, bushy, crested, or nearly naked; feet broad and short; toes subequal, with sharp curved claws; hind foot with a short, clawless, first toe, or hallux; soles of feet naked and granulated, with fine transversely granulated pads, that of the hallux being frequently divided into two. Pouch practically wanting; number of teats varying from two to five pairs. Three pairs of premolar teeth in each jaw, except in *P. cristicauda*, where the last lower one is wanting.

Although, as we have had occasion to mention in analogous instances, the name of Pouched Mice is far from being free from objection, yet since the scientific names of neither this nor the following genus lend themselves readily to conversion into English, we are compelled to use the colonial designation as the vernacular names of the members of both genera.

The pretty little animals belonging to the present genus thus designated, range over the whole of Australia and New Guinea, together with the adjacent islands, and are completely arboreal and inscetivorous in their habits. As is well stated by Mr. Thomas, they appear in the Australasian region to occupy the place held in India and the adjacent countries by the Tree-

Shrews (*Tupaia*), and in South America by the smaller kinds of Opossum. The largest of the thirteen known species does not exceed a Common Rat in size, while the majority are considerably smaller than that animal. Whereas in all the Australian species the fur of the back is not striped, the majority of the Papuan representatives of the genus have striped backs, although a few agree in this respect with their Australian congeners.

I. CRESTED-TAILED POUCHED MOUSE. PHASCOLOGALE CRISTICAUDA.

Chætocercus cristicauda, Krefft, Proc. Zool. Soc., 1866, p. 435.
Phascologale cristicaudata, Thomas, Ann. Mus. Genova, ser., 2, vol. iv., p. 509 (1887); *id*. Cat. Marsup. Brit. Mus., p. 276 (1888).

Characters.—Size medium; fur close and soft. General colour sandy brown, becoming paler on the under-parts. Ears short, rounded, and very broad; tail short, with a prominent crest of black hairs, lengthening towards the tip, on the upper surface of the terminal half. Last lower premolar tooth absent in the type specimen. Length of head and body about 5 inches; of tail 3½ inches.

Distribution.—South Australia.

Although closely allied to the next, this little-known species may be distinguished therefrom by its crested tail and more uniform coloration.

II. FRECKLED POUCHED MOUSE. PHASCOLOGALE APICALIS.

Phascologale apicalis, Gray, Ann. Mag. Nat. Hist., vol. ix., p. 518 (1842); Thomas, Cat. Marsup. Brit. Mus., p. 277 (1888).
Antechinus apicalis, Gray, List. Mamm. Brit. Mus., p. 99 (1843).

Characters.—Size medium; fur coarse. General colour freckled reddish-grey; under-parts dull white or yellowish; a whitish ring round each eye. Ears short and rounded, clothed on both surfaces with short grey hairs. Front and outside of fore leg rufous, remainder of outer surface of limbs, as well as the feet, grey; soles of feet granulated, the pad of the hallux rarely divided. Tail short, hairy, coloured above like the back, inferiorly grey or yellowish-grey, and the extreme tip black. Eight teats. Length of head and body about 5 inches; of tail 3½ inches.

Distribution.—West, and probably also North, Australia.

Habits.—This Pouched Mouse is found not uncommonly in the neighbourhood of the Swan River and King George's Sound, and is known to the settlers of the last-named district by the name of the "Dibbler." A correspondent of Gould, who was fortunate enough to find a female with young, states that these were seven in number, and quite naked and blind. He observes that above the teats of the mother there "is a very small fold of skin, from which the long hairs of the under-surface spread downwards, and effectually cover and protect the young. This fold in the skin of the abdomen is the only approximation to a pouch which I have found in any of the members of the genus. The young are very tenacious of life; those mentioned above lived nearly two days attached to the mammæ of the dead mother."

III. CHESTNUT-NECKED POUCHED MOUSE. PHASCOLOGALE THORBECKIANA.

Phascologale melas, Schlegel and Müller, Verhandl. Nat. Ges., p. 149 (1839-44).

Phascologale thorbeckiana, Schlegel, Ned. Tijdschr. Dierk., vol. iii., p. 257 (1866); Thomas, Cat. Marsup. Brit. Mus., p. 278 (1888).

Characters.—Size large; form comparatively stout; fur coarse and harsh, with the under-fur very thin. General colour richly variegated chestnut, black, and yellow; back with three black stripes; head dark yellowish-rufous, with a median black stripe commencing on the muzzle and passing backwards along the neck and back to the rump. Ears small, thinly covered at the back with black hairs. Crown and back of head, as well as top and sides of neck (save for the median stripe) rich chestnut-red. Chin and chest pale rufous; remainder of underparts yellowish-grey. Fore legs rich rufous like the neck, the hind ones darker; fore paws brown, the hinder brown or yellowish-brown; soles of hind feet naked, with five pads, that on the hallux being, at most, indistinctly divided. Tail evenly tapering, long-haired above and on the sides, and short-haired beneath; the long upper hairs maroon-red, like the rump; upper surface of the tip black, and the under side rufous or brownish. Six teats. The last premolar tooth wanting in some specimens. The length of the head and body rather less than in *P. penicillata* (*infrà*, p. 145).

Distribution.—North-western New Guinea.

This species received its earliest name of *P. melas* owing to its having been described from a black or melanistic individual, and since, as Mr. Thomas remarks, it is by far the most brilliantly coloured member of the family, if not indeed of the entire order, such a misleading title is clearly inadmissible. Together with the allied species, it affords a striking example of the characteristic brilliant coloration of Papuan animals as contrasted with their allies in other parts of the world.

IV. RED-TAILED POUCHED MOUSE. PHASCOLOGALE WALLACII.

Myoictis wallacei, Gray, Proc. Zool. Soc., 1858, p. 112.

Phascologale wallacei, Schlegel, Ned. Tijdschr. Dierk., vol. iii., p. 355 (1866); Thomas, Cat. Marsup. Brit. Mus., p. 280 (1888).

Phascologale (Chætocercus) pilicauda, Peters and Doria, Ann. Mus. Genova, vol. xvi., p. 668 (1881).

Characters.—Rather smaller than the last, from which it may be distinguished by the bright rufous neck and tail, the divided pad of the hallux in the hind foot, and by the median black stripe on the head being either indistinct or wanting, as well as by its generally less brilliant coloration, as especially shown by the paler head and neck. The tail is more bushy; and the last premolar tooth in both jaws considerably larger.

Distribution.—South-western New Guinea and the Aru Islands.

V. ORANGE-BELLIED POUCHED MOUSE. PHASCOLOGALE DORIÆ.

Phascologale doriæ, Thomas, Ann. Mus. Genova, ser. 2, vol. iv., p. 208 (1886); *id.* Cat. Marsup. Brit. Mus., p. 282 (1888).

Characters.—Size medium; fur thick, close and soft, with abundant dark slaty-grey under-fur; head long and slender. General colour dark grizzled orange-brown, the hairs on the back being tipped with orange; back with a single well-defined black stripe, commencing on the back of the head; underparts dull orange rufous. Tail comparatively short-haired, black, with the extreme tip white. Four teats.

Distribution.—North-western New Guinea.

VI. CHESTNUT-BELLIED POUCHED MOUSE. PHASCOLOGALE DORSALIS.

Phascologale dorsalis, Peters and Doria, Ann. Mus. Genova, vol. viii., p. 335 (1876); Thomas, Cat. Marsup. Brit. Mus., p. 283 (1888).

Characters.—Agreeing with the preceding in the single-striped back, the character of the tail, and the presence of only four teats, this Papuan species may be distinguished by the hairs of the back being tipped with white, the rich chestnut hue of the under-parts, and the inferior length of the hind foot.

Distribution.—North-western New Guinea.

VII. SWAINSON'S POUCHED MOUSE. PHASCOLOGALE SWAINSONI.

Phascologale swainsoni, Waterhouse, Ann. Mag. Nat. Hist., ser. 2, vol. iv., p. 299 (1840); Thomas, Cat. Marsup. Brit. Mus., p. 285 (1888).

Antechinus swainsoni, Gould, Mamm. Australia, vol. i., pl. xxiv. (1854).

Antechinus niger, et A. moorei, Higgins and Petterd, Proc. Roy. Soc. Tasmania, 1882, p. 172, and 1883, p. 182.

With this species we revert to the non-striped representatives of the genus, all of which, with the exception of one species from the Aru Islands, are Australian, one of the latter ranging into New Guinea.

Characters.—Size medium; fur very long, soft and thick. General colour deep rufous or umber-brown, under-parts dull brownish-grey; muzzle long; ears short and broad, covered with short dark brown hairs; feet dark brown; soles of the hinder pair with five pads, that of the hallux being at most but indistinctly divided; front claws very long and strong. Tail moderate, short-haired, uniformly dark brown. Probably ten teats. Length of head and body about 5 inches; tail of 4 inches.

Distribution.—Tasmania and South-eastern Victoria.

VIII. LITTLE POUCHED MOUSE. PHASCOLOGALE MINIMA.

Dasyurus minimus, Geoffr., Ann. Muséum, vol. iii., p. 362 (1804).

Phascologale minima, Temminck, Monogr. Mamm., vol. i., p. 59 (1827); Thomas, Cat. Marsup. Brit. Mus., p. 287 (1888).

Antechinus minimus, Gray, List Mamm. Brit. Mus., p. 99 (1843).

Antechinus rolandensis, et A. concinnus, Higgins and Petterd, Proc. Roy. Soc. Tasmania, 1882, p. 172, and 1883, p. 184.

Characters.—Size medium; form Mouse-like; fur thick, close and rather harsh. General colour grey, suffused with yellow or rufous, especially on the hinder-parts; under-parts, inclusive of chin, dirty yellowish-grey. Ears short, nearly naked, generally tufted at the base. A yellow patch on the front and outer side of the hip. Feet naked, yellow, or brown; soles of hind feet naked, usually with five pads, but that of the hallux occasionally divided; claws very long and strong. Tail short, closely short-haired, brown above, paler below. Number of teats unknown. Length of head and body about 5½ inches; of tail 3½ inches.

Distribution.—Tasmania and adjacent islands.

IX. YELLOW-FOOTED POUCHED MOUSE. PHASCOLOGALE FLAVIPES.

Phascologale flavipes, Waterhouse, Proc. Zool. Soc., 1837, p. 75; Thomas, Cat. Marsup. Brit Mus., p. 289 (1888).

Phascologale rufogaster, Gray, in Grey's Australia, Appendix, vol. ii., p. 407 (1841).

Antechinus stuarti, Macleay, Ann. Mag. Nat. Hist., vol. viii., p. 241 (1841).

Antechinus flavipes, Gray, List Mamm. Brit. Mus., p. 99 (1843).

(*Plate XXVII.*)

Characters.—Size small or medium; form stout; fur close

PLATE XXVII

YELLOW-FOOTED POUCHED-MOUSE.

and rather crisp General colour grey, suffused with yellow or rufous; under-parts yellow or rufous. Ears rather large, naked above and tufted externally at the base with yellow or grey hairs. Limbs coloured more or less like the under-parts; soles of hind feet naked, with six pads, that of the hallux being usually divided; claws small and delicate. Tail short-haired, brown or yellow above, paler beneath, the terminal inch black in some examples. Eight teats. Length of head and body about 5 inches; of tail 3½ inches.

Distribution.—From New Guinea throughout Eastern Australia to South Australia.

Variety.—Replaced in Western and Northern Australia by a variety (*P. leucogaster* of Gray) distinguished from the typical form by the nearly or quite pure white hue of the under-parts and limbs, in which the bases of the hairs still, however, retain the normal slaty grey tint.

It may be mentioned that Mr. Thomas gives the distribution of this species as exclusively Australian, and that its extension to New Guinea is added on the authority of Mr. Ogilby.

In its Australian haunts this species, according to Gould, may frequently be seen both on the ground and in trees. When on the latter, it clings very closely to the bark, keeping its legs far apart, and moving with a series of sudden little starts, somewhat after the manner of a Tree-creeper.

Krefft says that this lively little animal is the most abundant of its tribe, "and, though nocturnal, is often seen during the daytime. It used to be so common near the camp on the Murray that I have often captured several specimens whenever a load of wood was brought in. I kept many alive, and always found that, like the other species of *Phascologale*, it would attack and kill any number of Mice, if put into the same box. The shallow pouch of the female is provided with ten teats, and as many young are sometimes attached to them."

X. PIGMY POUCHED MOUSE. PHASCOLOGALE MINUTISSIMA.

Antechinus minutissimus, Gould, Proc. Zool. Soc., 1851, p. 284.

Antechinus maculatus, Gould, *loc. cit.*

Phascologale maculata, Wagner, in Schreber's Säugeth, Suppl., vol. v., p. 202 (1855).

Phascologale minutissima, Wagner, *op. cit.*, p. 203; Thomas, Cat. Marsup. Brit. Mus., p. 292 (1888).

Characters.—Size very small; fur soft, short, and fine, consisting chiefly of under-fur. General colour finely-grizzled mouse-grey; under-parts somewhat paler. Ears medium, thinly clothed with short hairs. Feet pale brown; soles of hind feet naked, with seven pads; tail of moderate length, short-haired. Pouch fairly well developed; eight teats. Length of head and body about three inches; of tail 2½ inches.

Distribution.—Central and Southern Queensland.

This pretty little animal is the smallest representative of the genus, and, both in form and coloration, simulates the common House-Mouse to a remarkable degree.

XI. LONG-TAILED POUCHED MOUSE. PHASCOLOGALE LONGICAUDATA.

Phascologale longicaudata, Schlegel, Ned. Tijdsehr. Dierk., vol. iii., p. 356 (1866); Thomas, Cat. Marsup. Brit. Mus., p. 293 (1888).

Characters.—May be distinguished from the last by being nearly double the size, and by the length of the tail exceeding that of the head and body; the latter feature distinguishing it from all the other species with the exception of *P. calura.* The nearest approach to this condition is made by *P. wallacii,* in which the length of the tail is equal to that of the head and body.

Distribution.—Aru Islands.

PLATE XXVIII

BRUSH-TAILED POUCHED-MOUSE

XII. BRUSH-TAILED POUCHED MOUSE. PHASCOLOGALE PENICILLATA.

Didelphis penicillata, Shaw, Gen. Zool., vol. i., pt. 2, p. 502 (1800).
Dasyurus penicillatus, Geoffr., Ann. Muséum, vol. iii., p. 361 (1804),
Dasyurus tafa, Geoffr., *op. cit.,* p. 360.
Phascologale penicillata, Temminck, Monogr. Mamm., vol. i., p. 58 (1827); Thomas, Cat. Marsup. Brit. Mus., p. 294 (1888).

(*Plate XXVIII.*)

The present and following species may be distinguished from all the other members of the genus with uniformly coloured backs by the tip of the tail being evenly tufted all round, whereas in the others the tail is either short-haired, or crested on the terminal portion of the upper surface only.

Characters.—Size large; form stout and strong; fur short and coarse. General colour finely-grizzled pale grey; under-parts and inner surface of limbs white or pale grey, the hairs on the pouch being rufous; muzzle with an indistinct darker stripe. Ears very large, nearly naked. Feet grey; soles of hind feet with the five main pads much elongated, the pad of the hallux undivided, and a minute additional pad on the outer margin of the hinder part; claws long and strong. Tail long and thick, its terminal half or three-fifths uniformly covered on all sides with long black hairs, forming a conspicuous brush. Ten teats. Length of head and body about 10 inches; of tail 6 inches.

Distribution.—All Australia, with the exception of the extreme north.

Habits.—The present species, which may be compared in size to an English Squirrel, has an unusually wide range.

Dwelling like its allies in trees, it makes its nest in the hollows of their trunks and branches, and feeds chiefly upon insects. In some districts it is so bold as to enter the houses of the colonists, by whom (whether justly or unjustly, we know not) it is accused of killing their poultry. From the large number of its teats, this species is doubtless a very prolific animal, although it does not appear to be ascertained whether it breeds more than once during the year.

A widely distributed species, it is found, according to Krefft, occasionally even in the neighbourhood of Sydney, and extends its range right across the Continent to the west coast; it is, however, very rare in the neighbourhood of the Murray river. In its general habits it is probably very similar to the next species, which is generally found in the hollow stems of trees. Of some specimens kept in confinement, Krefft writes that their movements were Cat-like, but very graceful; and that the animal resembled all the other members of the genus in being strictly nocturnal. On one specimen of *P. calura*, with ten teats, eight young ones were observed tightly clinging thereto, being concealed by the long hair on the under surface of their parent's body.

XIII. LESSER BRUSH-TAILED POUCHED MOUSE. PHASCOLOGALE CALURA.

Phascologale calura, Gould, Proc. Zool. Soc., 1844, p. 104; Thomas, Cat. Marsup. Brit. Mus., p. 296 (1888).

Characters.—Size medium; form slender; fur long, soft and fine. General colour grey, with a faint rufous tinge; underparts white Ears very large, almost naked, with well-marked tufts of red hair at the base. Feet white; soles of hind feet with five pads, that of the hallux being elongated, but undivided; claws small and weak. Tail long, the basal half short-haired, rufous above and dark brown beneath, the terminal half black and slightly bushy all round. Number of teats not known. Length of head and body about 5 inches; of tail 6 inches.

In addition to its inferior dimensions, this species may be readily distinguished by the base of the tail being red instead of grey, as well as by the greater proportionate length of that appendage.

Distribution.—South and West Australia.

THE NARROW-FOOTED POUCHED MICE. GENUS SMINTHOPSIS.

Sminthopsis, Thomas, Ann. Mus. Genova, ser. 2, vol. iv., p. 503 (1887).

Body unspotted; form slender and graceful; ears large and broad; tail moderate or short, short-haired, sometimes thickened; hind feet slender and delicate, with subequal toes furnished with small delicate claws; a clawless first toe, or hallux, on the hind foot; soles of hind feet partially haired, the naked portion granulated, and with or without pads, the latter, when present, being either smooth or but faintly striated. Pouch well-developed; eight or ten teats. Three pairs of premolar teeth in each jaw.

The Pouched Mice of this genus may be readily distinguished from *Phascologale* by the conformation of their hind feet, which are narrow, with granulated or hairy soles; whereas in the latter these feet are broad, with smooth naked soles.

The Narrow-footed Pouched Mice, which are confined to Australia and Tasmania, although agreeing with the members of the preceding genus in being insectivorous, differ in being exclusively terrestrial. Hence they may be likened to the Shrews of other regions. The three smaller species are so alike that their discrimination is a matter of some difficulty; and it is stated that a larger series of specimens than is now available in English collections is required before they can be fully defined.

I. THICK-TAILED POUCHED MOUSE.   SMINTHOPSIS CRASSICAUDATA.

Phascologale crassicaudata, Gould, Proc. Zool. Soc., 1844, p. 105.
Podabrus crassicaudatus, Gould, Mamm. Australia, pl. xlvii. (1845).
Podabrus macrurus, Gould, Proc. Zool. Soc. 1845, p. 79.
Antechinus (Podabrus) froggatti, Ramsay, Proc. Linn. Soc. N. South Wales, ser. 2, vol. ii., p. 552 (1887).
Sminthopsis crassicaudata, Thomas, Cat. Marsup. Brit. Mus., p. 306 (1888).

Characters.—Size small; form light and delicate; fur very soft and fine. General colour clear ashy grey; chin white; under-parts greyish-white. Ears very large and rather pointed, the anterior portion of their backs dark brown, forming a marked contrast with the lighter posterior part. Feet white. Soles of hind feet clothed for the greater portion of their extent with velvety hairs, the naked parts granulated, without distinct striated pads. Tail short, thickened, tapering, grey above, and white beneath. Ten teats. Length of head and body about 3½ inches; of tail 2 inches.

Distribution.—All Australia except the extreme north.

Habits.—According to Krefft, this pretty little species breeds in July and August, the female producing from six to nine young, which are placed in the shallow pouch. Occasionally, however, as many as ten are born in a litter. At times the creature utters a kind of hoarse screech; but except for this is silent.

In confinement it appears to thrive fairly well; but if two or more are left together, they are almost sure to fall out, and

PLATE XXIX

COMMON POUCHED MOUSE

seriously damage or kill one another. Like their allies, they are inveterate enemies to Mice, which they fall upon at once with great ferocity, killing more than they can possibly devour.

When treating of this species, Krefft makes an interesting remark relating to a peculiar habit of the family in general. "A singular peculiarity," he writes, "in all the *Dasyuridæ* is that they carry their ears folded down, never erect, when alive; and, though I do not want to find fault with Gould's beautiful work, I must say that, in this respect, the representations he gives of this tribe of the animals of Australia, are not over true to nature." We fear that the same remark would apply to the plates with which the present work is illustrated.

II. COMMON POUCHED MOUSE. SMINTHOPSIS MURINA.

Phascologale murina, Waterhouse, Proc. Zool. Soc., 1837, p. 76.
Phascologale albipes, Waterhouse, *op. cit.*, 1842, p. 48.
Antechinus fuliginosus, Gould, Mamm. Australia, vol. i., pl. xli., 1852.
Antechinus albipes, Gould, *op. cit.*, pl. xlii.
Antechinus murinus, Gould, *op. cit.*, pl. xliii.
Sminthopsis murina, Thomas, Cat. Marsup. Brit. Mus., p. 303 (1888).

(*Plate XXIX.*)

Characters.—Size small; form very slender and delicate; fur soft and fine. General colour finely grizzled Mouse-grey; chin white; under-parts greyish-white. Ears variable in size, their backs uniform slaty flesh-colour. Feet, and sometimes also the fore legs, white. Greater portion of soles of hind feet naked and finely granulated, without distinct striated pads. Tail moderate, slender, not thickened, brown above, and grey or white beneath. Eight teats. Length of head and body about $3\frac{1}{2}$ inches; of tail nearly the same.

Distribution.—Australia, South of the tropics.

III. WHITE-FOOTED POUCHED MOUSE. SMINTHOPSIS LEUCOPUS.

Phascologale leucopus, Gray, Ann. Mag. Nat. Hist., vol. x., p. 261 (1842).

Antechinus leucopus, Gray, List. Mamm. Brit. Mus., p. 100 (1843).

Antechinus ferrugineifrons, Gould, Mamm. Australia, vol. i., pl. xxvi. (1854).

Podabrus leucopus, et P. mitchelli, Krefft, Proc. Zool. Soc., 1866, p. 433.

Antechinus leucogenys, Higgins and Petterd, Proc. Roy. Soc. Tasmania, 1882, p. 172.

Characters.—Size medium; form slender; fur close, fine, and straight. General colour uniform dark greyish-brown or Mouse-colour, with no prominent markings; under-parts and feet white. Ears large and broad, uniform slaty-grey on the backs. Soles of fore feet finely granulated, with six pads; those of hind feet hairy behind and coarsely granulated in front, with four small and finely striated pads. Tail moderate, slender, shorter in specimens from the south than in those from the extreme north, grey or brown above, and white beneath. Length of head and body about 4 inches; of tail 3½ in southern, and 4 in northern forms. The number of teats unknown.

Distribution.—Eastern Australia, from Cape York to Tasmania.

IV. STRIPED-FACED POUCHED MOUSE. SMINTHOPSIS VIRGINIÆ.

Phascologale virginiæ, De Tarragon, Rev. Zool., 1847, p. 117.

Phascologale (Sminthopsis) virginiæ, Collett, Zool. Jahrbuch, vol. ii., p. 866 (1887).

Sminthopsis virginiæ, Thomas, Cat. Marsup. Brit. Mus., p. 300 (1888).

Characters.—Size large; fur rather short, very soft and silky. General colour grizzled grey; under-parts white or pale yellow; face sandy rufous with one median and a pair of lateral black stripes; cheeks, sides of neck, and tufts at bases of ears bright rufous. Ears very large, and nearly naked. Outer surface of upper part of fore legs and thighs coloured like the back; remainder of limbs white. Soles of hind feet probably as in the last species. Tail short-haired, dark brown above, paler beneath. Number of teats unknown. Length of head and body about 5 inches; of tail nearly the same.

Distribution.—Herbert River district, Queensland.

THE LONG-LEGGED POUCHED MICE. GENUS ANTECHINOMYS.

Antechinomys, Krefft, Proc. Zool. Soc., 1866, p. 434.

Body unspotted; ears very large; tail very long and tufted; limbs much elongated, the lower portions of the legs and the hind feet being disproportionately long; toes short and sub-equal; no first toe (hallux) to the hind foot; soles of feet without distinct pads, the greater portion of those of the hinder pair being hairy. Number of teats unknown. Three pairs of premolar teeth in both jaws; canine teeth very small.

The special characteristics of this genus are the great elongation of the limbs, and the absence of the first toe of the hind foot.

The one known species is a jumping terrestrial animal, bearing the same relationship to the Narrow-footed Pouched Mice as the Jumping Shrews of Africa present to the Tree-Shrews of India. Like its allies, it feeds exclusively on insects. Mr. Thomas remarks that the relationship of this genus to *Sminthopsis* is similar to that presented by the Jumping Mice (*Hapalotis*) of Australia to ordinary Mice (*Mus*); this saltatorial mode of progression having been doubtless developed in accordance with the exigencies of the arid sandy country inhabited alike by *Antechinomys* and *Hapalotis*.

I. JUMPING POUCHED MOUSE. ANTECHINOMYS LANIGER.

Phascologale lanigera, Gould, Mamm. Australia, vol. i., pl. xxxiii. (1856).
Antechinomys laniger, Krefft, Proc. Zool. Soc., 1866, p. 434; Thomas, Cat. Marsup. Brit. Mus., p. 309 (1888).

Characters.—Size small; form slender and graceful; fur long, soft, and fine. General colour slaty-grey; under-parts white; a fawn-coloured patch behind each ear. Ears suboval, almost entirely covered with short, fawn-coloured hairs. Lower portions of limbs, inclusive of the feet, white. Tail very long and slender, short-haired and fawn-coloured, except the terminal inch, which is tufted and black. Length of head and body about 3½ inches; of tail 5 inches.

Distribution.—The interior of New South Wales and Southern Queensland.

It may be mentioned that, through ignorance of their habits, a pair of these beautiful little animals are represented in Gould's " Mammals of Australia " as disporting themselves on the bough of a tree, whereas, as we have said, they are purely terrestrial. This plate, which is also by no means remarkable for the accuracy with which the animals are portrayed, has been reproduced as a woodcut in Brehm's " Thierleben."

THE MARSUPIAL ANT-EATERS. GENUS MYRMECOBIUS.

Myrmecobius, Waterhouse, Proc. Zool. Soc., 1836, p. 69.

The beautiful and curious little animal, which forms the sole representative of this genus, differs so remarkably from all the other members of the *Dasyuridæ*, that it is clearly entitled to form a sub-family (*Myrmecobiinæ*), if, indeed, as many zoologists think, it should not be referred to a family by itself. Regarding it as the representative of a sub-family only, the *Myrmecobiinæ* will be distinguished from the preceding sub-family of the *Dasyuridæ* by the following characteristics :—

Tongue long, cylindrical, and extensile ; nose naked, and grooved below ; lower lip pointed, and projecting beyond the teeth ; chest furnished with a complex gland, opening to the surface by means of several large and distinct apertures. Cheek-teeth small and delicate, the molars being more than four in number on each side of both the upper and lower jaw, and those of the latter with their inner cusps larger than the outer ones.

As a genus, *Myrmecobius* may de defined as follows :—

Form graceful and Squirrel-like ; ears long and narrow ; fore feet with five, and the hinder with four toes, the hallux being wanting externally ; all the toes furnished with long claws adapted for digging ; soles of fore feet partially, and those of hind pair entirely, naked, their pads small and granulated. Tail long and bushy. Pouch obsolete. Teats, according to Thomas, four in number, although there is a statement of Gould implying the existence of seven or eight. Three pairs of premolar teeth in each jaw, and five pairs of molars, with the occasional presence of a sixth pair in the lower jaw ; the total number of teeth, inclusive of the incisors and canines, thus being either 50 or 52. Very rarely a fourth pair of lower incisors occurs, bringing up the extreme total to 54.

The especial interest of this genus, in which the number of teeth is greater than in any other mammals which have the teeth differentiated into series, lies in its relationship to the extinct Marsupials of the Mesozoic rocks of Europe, as exemplified by the form and number of its molar teeth. Since these teeth are very remarkable, and unlike those of any other living members of the order, they merit especial notice. In the words of Waterhouse, "the true molars are very small, and longer than broad; those of the upper jaw presenting numerous small tubercles; in the lower jaw the outer edge of the molars is produced and divided by notches, so as to leave three or four bristly points in each. On the outer side of each molar," excepting the more anterior ones of the series, "are several small, blunt tubercles, which must be nearly on the level of the gum in the living animal. Between all the teeth, both of the upper and lower jaws, excepting the four posterior molars of the lower jaw, there is a space which is sometimes equal to the width of the teeth, but generally less. The ramus of the lower jaw is twisted in such a manner that the outer surface of the true molars comes in contact with the masticating surface of those of the lower jaw."

In addition to the similarity of the teeth, the genus shows a resemblance to the extinct Jurassic Marsupials of Europe in the presence of a narrow but well-defined channel, known as the mylo-hyoid groove, on the inner side of each branch, or ramus, of the lower jaw.

That the genus, like the fish *Ceratodus* of the rivers of Queensland, and several other peculiar Australian types, is a direct survivor from the Secondary fauna of Europe, there seems no reasonable doubt.

I. BANDED ANT-EATER. MYRMECOBIUS FASCIATUS.

Myrmecobius fasciatus, Waterhouse, Proc. Zool. Soc., 1836, p. 69; Thomas, Cat. Marsup. Brit. Mus., p. 312 (1888).

(*Plate XXX.*)

Characters.—Fur short, close, rough, and almost spiny. General colour bright rufous, grizzled on the head, and darkening posteriorly, where it is ornamented with a series of broad transverse white bands; a white stripe over each eye; under-parts clear pale yellow. Ears pointed, and clothed

PLATE XXX

BANDED ANTEATER

with close, short hairs; rufous on the backs, and yellowish internally. Claws dark horn-colour; third front toe shorter than the second and fourth; soles of fore feet with five small, round, finely granulated pads; those of the hind feet hairy, and with only three pads. Tail long-haired above, shorter haired beneath; the long upper hairs being grizzled yellow and black, and the short lower ones rich rufous. Length of head and body about 10 inches; of tail 7 inches.

Distribution.—South and West Australia.

Habits.—The first specimens known to Europeans of this beautiful little animal, which may be compared in size to an ordinary Squirrel, were obtained by an English subaltern—Mr. Dale,—during an exploring expedition in the interior of the country in the Swan River district. According to the original account, "two of these animals were seen within a few miles of each other; they were first observed on the ground, and on being pursued, both directed their flight to some hollow trees which were near. We succeeded in capturing one of them; the other was unfortunately burnt to death in our endeavour to dislodge it by fumigating the hollow tree in which it had taken refuge;—the country in which they were found abounding in decayed trees and ant-hills."

The specimen thus captured was brought to England and put into the hands of Waterhouse, by whom it was described under the name by which the animal is universally known. From the character of its teeth, coupled with its long, extensile tongue, and strong curved claws, this sagacious naturalist, taking into consideration the description of the country where the first example was captured, came to the conclusion that the creature lived on ants, which it first dug out from their nests, and then licked up with his tongue. This inference, it need scarcely be mentioned, has been fully confirmed by actual observation.

According to Gould, the Banded Ant-eater is, indeed, only to be found in localities where its favourite food is abundant. Mr. Gilbert, who had many oportunities of seeing this little animal in its native haunts, writes that "it appears very much like a Squirrel when running on the ground, which it does in successive leaps, with its tail a little elevated; every now and then raising its body, and resting on its hind feet. When alarmed, it generally takes to a dead tree lying on the ground, and before entering the hollow invariably raises itself on its hind feet, to ascertain the reality of approaching danger. In this kind of retreat it is easily captured, and, when caught, is so harmless and tame as scarcely to make any resistance, and never attempts to bite. When it has no chance of escaping from its place of refuge, it utters a sort of half-smothered grunt, apparently produced by a succession of hard breathings.

"The female is said to bring forth her young in a hole in the ground, or in a fallen tree, and to produce from five to nine in a litter. I have not myself observed more than seven young attached to the nipples. Like the members of the genus *Antechinus*, this animal has no pouch for the protection of the young; the only protection afforded their delicate offspring being the long hairs which clothe the under-surface of the abdomen of the mother."

It must be observed that in this very circumstantial account, which is from the pen of Mr. Gould's trusty and ill-fated collector, the statement as to the number of young is quite at variance with the presence of only four teats in the female. Unfortunately, there do not appear to be any recent observations on the number of young in a litter, and the attention of those who have an opportunity of seeing the animal in its native haunts may well be directed to this, as also to the number of teats of the female.

Although the Banded Ant-eater is chiefly a terrestrial creature, it is reported to ascend trees with facility, and to be partially arboreal in its habits; while it always frequents well-wooded districts. In capturing Ants, after the nest has been laid open by its fore paws, the long extensile tongue is protruded among the insects, and held there till a mass of them have collected upon it, when it is quickly retracted, and the food swallowed. The creature will, however, also eat insects of other kinds, and even, it is said, grass.

There have been recently discovered on the palate of the Banded Ant-eater traces of horny structures, which apparently correspond to the so-called horny teeth of the Duck-bill among the egg-laying Mammals. Since there is also a certain similarity between the molar teeth of these otherwise widely sundered Mammals, this is a fact of some interest, as tending to suggest a distant relationship between this primitive Marsupial and the still lower Monotremes.

THE MARSUPIAL MOLES. FAMILY NOTORYCTIDÆ.

Three pairs of incisor teeth in both jaws. Limbs subequal, short, stout, and strong, each with five distinct toes; in the hind foot a well-developed and clawed first toe, or hallux, not opposable to the others, which are unequal in size. No externally visible ear-conchs or eyes. Collar-bones, or clavicles, well-developed; and chevron-bones present on the inferior aspect of the vertebræ of the tail. Upper molar teeth triangular, with three cusps.

THE MARSUPIAL MOLES. GENUS NOTORYCTES.

Notoryctes, Stirling, Trans. Roy. Soc. South Australia, 1891, p. 154.

Upper aspect of muzzle covered with a hard horny shield, divided into two portions by a transverse ridge; mouth centrally placed; apertures of ears almost completely concealed by overhanging fur. The four inner toes of both fore and hind feet clawed, the fifth with a short, broad, horny nail; claws of the third and fourth front toes enormously enlarged; those of the corresponding hind toes curving outwards and backwards, and the toes of these feet decreasing in size from the second to the fifth; soles of both pairs of feet naked, and covered with tough, leathery, wrinkled skin, traversed in the hinder pair by oblique folds. Tail very short, hard, tough, and leathery, marked by conspicuous rings, thick at the base, but rapidly decreasing in size towards the extremity, which is blunt and knob-like. Pouch opening backwards; two minute teats. Two pairs of premolar teeth in each jaw, which, like the three pairs of incisor and single canine, are very minute. In the skeleton the five middle vertebræ of the neck are completely welded together, as are likewise certain vertebræ in the region of the haunch.

The single representative of this genus and family, which has been only recently made known to science, is a burrowing, Mole-like creature, standing quite alone among the Marsupials of Australia. That it is a primitive type is sufficiently indicated by the triangular, or tritubercular form of its upper molar teeth, which are quite unlike those of any other living member of the order, and likewise by the presence of chevron-bones beneath the vertebræ of the tail. In adaptation to its particular mode of life the creature is, however, in some respects markedly specialised.

Perhaps the most remarkable feature in connection with this animal is its curious resemblance, both as regards the structure of its molar teeth, its external form, and its mode of life, to the Golden Moles (*Chrysochloris*) of South Africa, which belong to the insectivorous order of placental Mammals. So marked, indeed, is this resemblance, that it must evidently be due either to parallel development induced by similar

modes of life, or to genetic affinity. Professor E. D. Cope, of Philadelphia, pronounced emphatically in favour of the latter view, at the same time expressing the opinion that *Notoryctes* was probably a member of the Insectivora, and not a Marsupial at all. Subsequent researches have, however, conclusively proved its Marsupial affinities; and since it is perfectly clear that the Marsupial Mole could not have been the ancestor of the Golden Moles, and the latter the original stock whence the whole of the other Insectivora took origin, there is probably no sort of relationship between the two genera, whose mutual resemblance would thus seem solely due to parallelism in development. At the same time it must be confessed that the exact similarity between the molar teeth of the two is somewhat difficult to explain, although it may probably be accounted for by both having retained this primitive type of tooth from early ancestors.

THE MARSUPIAL MOLE. NOTORYCTES TYPHLOPS.

Notoryctes typhlops, Stirling, Trans. Roy. Soc. South Australia, 1891, p. 154; Ogilby, Cat. Australian Mammals, p. 5 (1892).

(*Plate XXXI.*)

Characters.—Size small; form stout; fur long, soft, and of a bright lustrous silky appearance. General colour golden red, sometimes darker across the loins, and a patch of darkish red surrounding the pouch; inner surface of pouch sparsely lined with reddish fawn-coloured hairs. Upper surface of tail covered with fur similar to that of the back on its basal half, the sides and lower surface being naked. Length of head and body about 5 inches; of tail rather more than an inch.

Distribution.—Central South Australia.

Habits.—It appears from the account given by its describer Dr. Stirling that the Marsupial Mole, or "Ur-quamata," as it is called by the blacks, was first discovered by Mr. Coulthard, manager of some of the estates of the Willowie Pastoral Company, who, on reaching his camp one evening on the Finke river, was attracted by certain peculiar and unfamiliar tracks. On following these up he found a Mole-like animal lying beneath a tuft of porcupine-grass, or spinifex. Other specimens were subsequently obtained by Mr. Bishop in the same neighbourhood, distant about a thousand miles from Adelaide; the country consisting of flats and dunes of red sand, covered with porcupine-grass and acacias, and the rainfall being small. Most of the specimens have been captured by the natives, who track them after rain, when their trail is conspicuous in the sand. The creature seems to be on the move only in warm weather, so that the short rainy season in the summer appears to be the one most favourable for its capture. At such seasons it appears to be perpetually engaged in burrowing; "emerging from the sand, it travels on the surface for a few feet, at a slowish pace, with a peculiar sinuous motion, the belly much flattened against the ground while it rests on the outsides of its fore paws, which are thus doubled in under it. It leaves behind it a peculiar sinuous triple track, the outer impressions, more or less interrupted, being caused by the feet, and the central continuous line by the tail, which seems to be pressed down in the rear. . . . . It enters the sand obliquely and travels underground either for a few feet or for many yards, not apparently reaching a depth of more than two or three inches, for whilst underground its progress can often be detected by a slight cracking or moving of the surface over its position. In penetrating the soil, free use, as a borer, is made of the conical snout with its horny protecting shield, and the powerful scoop-like fore paws are also early brought into play. As it disappears from sight, the hind limbs, as well, are used to throw the sand backwards, which falls in again behind it as it goes, so that no permanent tunnel is left to mark its course. Again emerging, at some distance, it travels for a few feet upon the surface and then descends as before. I could hear nothing of its making, or occupying at any time, permanent burrows. Both my informants," continues Dr. Stirling, "lay great stress on the phenomenal rapidity with which it can burrow, as observed both in a state of nature, and in captivity." From this account the Marsupial Mole may be said to swim in the sand much in the same manner as a Porpoise or a Dolphin swims in the ocean, alternately disappearing for a short distance beneath the surface, and progressing with its body exposed. From the nature of the soil, and its periodical appearances on the surface, it will be obvious that the creature has not to perform work anything like as hard as that undergone by the Common Mole in the excavation of its underground tunnels. The reader will not fail to notice the beautiful adaptation of the creature to its surroundings as exemplified by the harmony existing between the coloration of its fur and that of the red sand of the desert.

Further information as to the habits of the Marsupial Mole was supplied to Dr. Stirling by Mr. Bishop from observations made on a specimen in captivity. The latter gentleman having a living example brought to him, he kept it, writes Dr. Stirling, "in a box of sand, in which was placed a tussock of porcupine-grass (*Triodia irritans*), so as to imitate its natural surroundings as nearly possible. At the same time precautions against exposure to cold were taken by covering up the box with blankets, and the sand in the box was frequently changed, the fresh supply being first warmed and moistened. It was fed on the 'witchetty' [a kind of grub] previously referred to, two or three small grubs or a single large one being given daily. These it ate with such evident eagerness and avidity as to suggest that the animal was accustomed to that kind of food.

Occasionally it was offered Beetles of a species that is found under the porcupine clumps, but though these disappeared from the box, it is not certain that they were actually eaten. Ants, also, were tried with a previous specimen, but it seemed as if it was the *Notoryctes* which ran the chance of being eaten. Strong support is afforded to the suggestion that the 'witchetty' forms a part, at least, of its diet by the fact that, as previously observed, acacias are plentiful in the sand-hills, which seem to be the natural haunt of the animal, and the larva in question are found in these roots at a depth of a foot or more. The suggestion is also confirmed by the statements of the natives and by the observations of Mr. Bishop himself, who found traces of underground burrowings around the stems of the acacias. Still it is not clear how the 'witchetties,' which are buried in the wood, are dislodged."

The writer then goes on to say that when inspected in its box, the Ur-quamata "would usually be found lying curled up in the sand, but not covered by it, and when the hand was put down immediately in front the little animal would climb into it and claw it all over. It seemed quick of hearing, and always awakened immediately on lifting the lid of the box. A very slight whistling noise was sometimes made while the specimen, kept so long in captivity, was burrowing about in the box; but it was not certain whether this was a respiratory or a true voice sound; and a previous specimen was heard, when held, to make a faint chirping like that of a newly-hatched chick. None of the other specimens, however, emitted any sound that was audible.

"Whenever the sand was changed by a fresh, warm, and moist supply the animal immediately commenced to burrow, and on warm sunny days when it was let out in the open it would, if the sand were hard, run a little way fairly quickly; but wherever the surface was soft it would begin burrowing directly, and, as soon as it had got a fair start, it progressed with great rapidity."

What becomes of the Ur-quamata during the cold season, when, according to native reports, it is no more seen above ground, has not yet been ascertained. From the circumstance that during a slight frost two specimens kept in a box of sand died, it may, however, be inferred that during the winter these little creatures must burrow to such a depth in the sand as to be beyond the reach of the slight frosts met with in the districts they inhabit, and that in such burrows they undergo a more or less complete hibernation. No sort of information has hitherto been ascertained as to the breeding habits of these little creatures, a young one not even having been seen.

It may be added, in conclusion, that there is no sort of support to the suggestion that the Marsupial Mole forms a connecting link between the Marsupials and the egg-laying Mammals. Among the former it is believed by Dr. Gadow to be more nearly related to the Opossums than to either of the other Australian families of Polyprotodonts.

THE OPOSSUMS. FAMILY DIDELPHYIDÆ.

With the Marsupial Mole we take leave of the Australian representatives of the order, and come to the single American family, which includes the whole of the remaining forms now living. Possessing the characters already given as distinctive of the Polyprotodont sub-division of the order, the Opossums may be characterised as a family as follows :—

Five pairs of upper, and three of lower incisor teeth. Hind feet with the four outer toes subequal and detached, and a well-developed, although clawless, inner toe, or hallux, which can be opposed to the other four. Tail generally very long, naked, scaly, and prehensile, but occasionally short and more or less hairy. Stomach simple ; intestine with a small or moderately-sized blind appendage, or cæcum. Pouch generally wanting, sometimes composed merely of two lateral folds of skin not united at the ends, and rarely complete. Three pairs of premolar teeth, the last of which always has a well-developed deciduous predecessor.

The Opossums, which include more than a score of species, varying in size from that of a Cat to that of a Mouse, are confined at the present day to America, where they range from the United States to Argentina. They may, however, be regarded as more especially characteristic of the wooded districts of the southern half of that continent, but during the early part of the Tertiary period their range included England and other parts of Europe. Very similar to one another in external appearance, as well as in internal structure, the whole of the species, with the exception of a single one distinguished by its webbed feet and aquatic habits, are included in one large genus, which, however, may be split up into several subgeneric groups, thus rendering it less unwieldy than would otherwise be the case. As a whole, the family is regarded by Mr. Thomas as very closely allied to the *Dasyuridæ;* one of its main claims to distinction being its isolated geographical distribution.

The Opossums being among the lowest of Marsupials, the existence of a complete pouch among certain species is of the greatest importance as showing that this organ is primitively an essential characteristic of Marsupials, and not one that has been specially developed to suit the exigencies of the various modes of life of the Australian members of the order. Consequently those forms, like the majority of the Opossums and the Banded Ant-eaters, which have either no pouch or merely a rudiment thereof, may be safely assumed to have lost that organ through specialisation. The reason of the loss of such an organ, which appears so admirably adapted for the protection of young born in the imperfect and helpless condition characteristic of all the Marsupials, is hard indeed to divine. In the case of the Opossums, the young of many species, after becoming detached from the teats, are enabled to travel about with her by twisting their prehensile tails round the tail of their parent; but it will be obvious that they are by no means so secure or well protected as they would be in a warm and comfortable pouch.

With the exception of the aquatic Yapock, or Water-Opossum, the members of the family are mainly arboreal animals, spending the hours of daylight concealed among the foliage or in the hollows of the trees they frequent, and only issuing forth at night to procure food. A few species belonging to the subgeneric group *Peramys* are, however, more or less terrestrial in their habits; and it is a remarkable fact that some of the normally arboreal species have extended their range to the open pampas of Argentina, where they have from necessity been forced to adopt a mode of life foreign to their nature, and thus affording an instance of the pliancy and adaptability of animals. Whereas the normally terrestrial species alluded to above resemble the Shrews both in appearance and mode of life, the majority of Opossums may be more aptly compared, as regards their *rôle* in nature, with the Tree-Shrews (*Tupaia*) of the Oriental region and the Broad-footed Pouched Mice of Australasia.

As regards their food, the smaller species feed mainly or exclusively on insects, which also constitute a considerable proportion of the nutriment of the larger kinds. These latter, however, also prey upon the smaller reptiles, as well as birds and their eggs; being, in fact, almost omnivorous. For the capture and comminution of insect prey the sharp teeth of these animals are especially adapted; the large number of cusps on the molars being just the structure best suited for piercing the hard wing-cases of Beetles. In their large number of incisor teeth—five pairs in the upper, and four in the lower jaw—the Opossums, as shown in the figure on page 129, differ from all other Mammals, the *Peramelidæ* coming next to them in this respect, with, at most, five pairs in the upper, and three in the lower jaw. These teeth are arranged in each jaw nearly in the form of a semicircle; in the upper jaw the inner pair are somewhat longer than the rest, from the nearest of which they are generally separated by a narrow interval, their form being nearly cylindrical, with a slight dilitation at the summit. The canine teeth are well-developed, those of the upper jaw being somewhat longer than those opposed to them.

In general form and appearance the majority of Opossums may be compared to the common Grey Rat, except that the muzzle is more elongated, and completely naked at the extremity, where it is perforated by the nostrils. In the larger species the form of the body is proportionately stouter. A peculiarity of the Opossums is that the teats, which are always numerous, varying from five to twenty-five, are always an odd number; either a single odd one, or as many as five being placed in the centre of a circle or oval formed by the others.

In the case of the short tailed group, forming the sub-genus *Peramys*, the series of specimens available to the author of the British Museum Catalogue of Marsupials was insufficient to allow of the proper determination of all the species, which may consequently stand in need of revision.

THE TRUE OPOSSUMS. GENUS DIDELPHYS.

Didelphys, Linn., Syst. Nat., ed. 10, vol. i., p. 54 (1760).

1. COMMON OPOSSUM. DIDELPHYS MARSUPIALIS.

Didelphys marsupialis, Linn., Syst. Nat., ed. 10, vol. i., p. 54 (1760); Thomas, Cat. Marsup. Brit. Mus. p. 323 (1888).

AZARA'S OPOSSUM

Didelphis karkinophaga, Zimmermann, Geogr. Geschichte, vol. ii., p. 226 (1780).
Didelphis carcinophaga, Boddaert, Elenchus Anim., vol. i., p. 77 (1785).
Didelphys cancrivora, Gmelin, Syst. Nat., vol. i., p. 108 (1789).
Didelphys virginiana, Kerr, Linn., Anim. Kingd., p. 193 (1792).
Philander virginiana, Tiedemann, Zool., p. 427 (1808).
Didelphys aurita, Wied, Beiträg. Nat. Brasil, vol. ii., p. 395 (1826).
Didelphys breviceps, Rennett, Proc. Zool. Soc., 1833, p. 40.
Variety—*Azara's Opossum*.
Didelphys azaræ, Temminck, Monogr. Mamm., vol. i., p. 30 (1827).
Didelphys albiventris, Lund, Blik. Bras. Dyr. Dansk. Afh., vol. viii., p. 127 (1841).
Didelphys pæcilotis, Wagner, Archiv. für Nat., vol. viii., p. 358 (1842).
Didelphys leucotis, Wagner, Abhandl. Akad. München, vol. v., p. 127 (1847).

(*Plate XXXII.*)

Characters.—This species, with its varieties, alone represents the typical group of the genus, which may be characterised as follows: Tail very long, scaly, naked, and markedly prehensile; size large; fur with long bristle-like hairs intermixed; fifth hind toe much shorter than the second, third, and fourth, which are of nearly equal length.

As distinctive characters of the typical variety, the following may be given:—

Size from three to five times that of any other Opossum; fur long, coarse, and thick, composed of a short, soft under-fur extending uniformly over the body, and a longer upper-fur consisting of coarse elongated bristles, chiefly confined to the upper-parts, more numerous along the middle line of the back than elsewhere. General colour varying from white to black, through all the intermediate mixtures, the individual bristly hairs being either white, black, or parti-coloured, but the under-fur invariably white at the base; colour of face also variable in its proportions of black and white, the northern forms having the face nearly wholly white, save for a darker streak running through each eye, and another on the crown of the head, but southern examples altogether much darker, and frequently black; underparts dirty white, often with black tips to the hairs. The naked muzzle broad, with a single median vertical groove, and two lateral notches in the upper lip. Ears large, leaf-like, and oval, varying in colour from black to white. Limbs brown or black, always darker than the body; fore feet with six, hinder with five pads, which are large and coarsely striated, although not divided. Pouch large and well-developed; teats varying from 5 to 13 in number. Tail haired for from one to four inches at the base like the body, elsewhere scaly and nearly naked. Length of head and body, in large specimens, about 18½ inches; of tail 17 inches.

Distribution.—America, from the United States to Chili and South Brazil; also Argentina.

Variety.—The variety known as Azara's Opossum (Plate XXXII.), which is equally as variable in coloration as the typical form, is distinguishable, according to Mr. Thomas, "by the prominence and sharp definition of the darker marks on the face, these forming strongly contrasted black stripes on a white ground, one running through each eye, and a median one passing from between the eyes backwards on to the crown and back of the neck. Specimens with these markings are generally smaller in size, and have, as a rule, white or parti-coloured ears, more hairy tails, and smaller teeth than the ordinary form, but no one of these characters is entirely constant."

Distribution.—Colombia, Ecuador, Peru, Bolivia, Chili, Paraguay, and South Brazil; that is to say, the countries lying to the west and south of the Amazonian region. Also Argentina and Patagonia (Hudson). Azara's Opossum is the only one which can be allowed to rank as a distinct geographical variety of the common Opossum, all the other variations to which distinct specific names have been applied having no sort of constancy, either in the form of the markings, or in geographical distribution.

Habits.—The common Opossum is a thoroughly arboreal species, and is chiefly noticeable for the large size of its body, which may be compared to that of a Cat, and for the full development of the pouch; the numerous young resorting to the latter until they attain considerable dimensions. An expert climber, the Opossum hunts eagerly among the boughs for birds and their nests, as well as for the smaller reptiles and larger insects; while it is reported to be very destructive to poultry. In climbing, it is assisted by its prehensile tail; and when pursued and wounded, displays the death-feigning instinct, suffering unbounded ill-treatment without moving a muscle or displaying a sign of life. Like many of the lower Mammals, its vital tenacity is wonderfully developed. The number of young at a birth may even amount to a dozen; and at the time these are brought forth the female makes a nest of dry grass at the root of a tree, or in some thick bush. When first born, the shapeless, naked young are extremely minute, and it seems marvellous how they are able to attach themselves to the teats in the pouch. They develop, however, with great rapidity, and in a short time attain the size of a Mouse, when they are able to leave the pouch, to which they return for shelter at the approach of danger, or for the purpose of obtaining nutriment.

The *Manicou*, as this species is called in some parts of

America, is one of the fifty-two Mammals inhabiting the island of Trinidad.

II. QUICA OPOSSUM. DIDELPHYS OPOSSUM.

Didelphys opossum, Linn. Syst. Nat., ed. 12, vol. i., p. 55 (1760); Thomas, Cat. Marsup. Brit. Mus., p. 329 (1888).
Didelphys quica, Temminck, Monogr. Mamm., vol. i., p. 36 (1827).
Metachirus opossum, et M. quica, Burmeister, Erläut. Faun. Brasil., pp. 69, 70 (1856).

Characters.—This species and the two following forms constitute the subgenus *Metachirus*, of which the characters are as follows :—

Size medium; toes of hind feet with the same relative lengths as in the preceding species, that is to say with the three middle toes subequal, and considerably exceeding the outermost one in length; pouch either rudimentary or well-developed; fur short and straight, without an admixture of bristles among the soft hairs.

The present species may be distinguished from its two nearest allies as follows :—

Size medium; fur short, straight, and somewhat crisp. General colour shining slaty-grey, of variable shades, but darker on the head, and lighter on the flanks than elsewhere; underparts yellowish, greyish, or white; face dark brown or black, with a pair of prominent white spots above the eyes, sometimes separated from another by a mere line. Muzzle long and slender; upper lip with a single pair of lateral notches. Ears large, leaf-like, rounded, and very thin; white at the base and black at the tip, with frequently a white spot on the head behind each of them. Pouch well-developed. Seven teats. Feet brown or brown and white; their pads large and rounded, those of the first toe on each foot being more or less distinctly

divided. Tail hairy for the first two or three inches at the base, then changing somewhat suddenly to scaly, and the tip gradually becoming white. Length of head and body about 10¾ inches; of tail slightly less.

Distribution.—Mexico to the Argentina.

Habits.—This animal, which is somewhat larger than a common Squirrel, is very common in many parts of Brazil, where it is known as the *Quica*. Feeding upon small birds, insects, and fruits, it passes its time, like the others of its kin, in sleeping, rolled up like a ball, during the day, and sallies forth to feed at night. The full development of the pouch indicates that the young have habits similar to those of the common species. By the French, the species is termed *Le Didelphe quatre-œil*, the conspicuous white spots on the forehead giving, at a distance, the appearance of a second pair of eyes.

III. RAT-TAILED OPOSSUM. DIDELPHYS NUDICAUDATA.

Didelphys nudicaudata, Geoffr., Cat. Mus., p. 142 (1803); Thomas, Cat. Marsup. Brit. Mus., p. 332 (1888).
Didelphys myosurus, Temminck, Monogr. Mamm., vol. i., p. 38 (1827).
Metachirus myosurus, Burmeister, Erläut. Faun. Brasil., p. 69 (1856).

(*Plate XXXIII.*)

Characters.—Size scarcely equal to that of the last, but the form more slender, and the limbs and tail relatively much longer; fur, very short, straight and crisp. General colour greyish-brown, more or less tinged, especially on the flanks, with yellowish or rufous; under-parts yellowish-white, sharply defined from the dark area above by a more or less well-marked yellow line; face brown or rufous-brown, darker round the eyes, and having a prominent white or pale yellow spot above each of the latter, which is, however, much smaller than in the preceding species. Muzzle with one pair of lateral notches in the lip; ears very large, broad, rounded, naked, transparent, and of a uniform slaty-grey colour. Pouch rudimentary or wanting. Nine teats. Front of limbs, inner side of lower part of hind legs, and feet pale brown. Feet longer and narrower than in the last, with their pads very distinctly defined, and those of the first toe in each foot generally undivided; the outermost hind toe scarcely reaching to the middle of the second joint of the fourth. Tail of great length, with its basal inch alone haired, elsewhere naked, save for a few short hairs between the scales, and its general brown colour changing to white at the tip. Length of head and body about 9½ inches; of tail 11¾ inches.

Distribution.—Costa Rica to Brazil.

The distinctive features of this species, which is very abundant in Brazil, and especially at Bahia, are the small spots above the eyes, coupled with the great length of the tail and the shortness of its furred portion. The absence of a pouch in a species otherwise so closely allied to *D. opossum* is very remarkable, indicating differences in the habits of the young, and also showing that the present is the more specialised form of the two.

IV. THICK-TAILED OPOSSUM. DIDELPHYS CRASSICAUDATA.

Didelphys crassicaudata, Desmarest, Nouv. Dict. d'Hist., vol. xxiv., table, p. 19 (1804); Thomas, Cat. Marsup. Brit. Mus., p. 334 (1888).

Didelphis macroura, Illiger. Abhandl. Akad. Berlin, 1811, p. 107.

Metachirus crassicaudatus, Hensel, Abhandl. Akad. Berlin, 1872, p. 121.

Didelphys turneri, Günther, Ann. Mag. Nat. Hist., ser 5, vol. iv., p. 108 (1879).

Characters.—Size as in *D. opossum*; form long, low, and remarkably Weasel-like; fur straight, thick, and soft. General colour rich soft yellow, becoming greyish down the middle of the back, and brighter on the flanks and under-parts; face like the back, without eye-spots or other conspicuous markings, but some of the hairs pencilled with brown. Upper edge of naked portion of muzzle with an upward rounded projection, sharply differentiated from the hair of the face. Ears very short and rounded, scarcely appearing above the level of the fur, the inner edge with a long conical projection at the base, and their substance thick, fleshy, and covered, except on the margins, with dark yellow hairs. Pouch wanting. Teats nine. Limbs very short, coloured like the body, but the feet browner; feet short, with the pads small and narrow, and the outermost hind toe reaching only to the middle of the first joint of the fourth. Tail extremely thickened at the base, where it appears to pass imperceptibly into the body; its basal half haired like the body, elsewhere short-haired, except for some two inches of the lower surface of the tip, where it is naked; brown or black at the base, and white at the tip. Size very variable; the length of head and body in one specimen about $10\frac{1}{4}$ inches, and of tail $8\frac{1}{3}$ inches; while in another the corresponding dimensions are $15\frac{1}{2}$ and $10\frac{1}{2}$ inches.

Distribution.—Guiana and South Brazil, unknown in the intermediate districts, and Argentina.

This species may be easily recognised by its Weasel-like form, uniformly coloured head, short ears, and thick hairy tail. The variation in size is very remarkable, being so great that some specimens are double the bulk of others of the same sex; the Brazilian examples being, according to Mr. Thomas, generally much larger than those from Guiana. *Didelphys turneri* was founded on an abnormal specimen with only four pairs of upper incisor teeth.

Habits—According to Mr. W. H. Hudson an Opossum which he identifies with *D. aurita* (= *D. marsupialis*), but which from its bright yellow colour, both above and below, and Weasel-like form, would appear to be the present species, ranges southwards into the Argentine pampas. There he describes it as both terrestrial and aquatic in its habits, frequenting the low-lying lands subject to inundation, and mostly devoid of trees. On dry land its habits are compared to those of a Weasel; while it dives and swims with ease in the small *lagunas* (lagoons) dotted over the pampas, constructing a globular nest of grass suspended from the flags and rushes which abound in such spots.

The same author states that Azara's Opossum likewise inhabits not only the pampas of Buenos Ayres, but also the desert regions of Patagonia, where not a tree is to be seen. There it shuffles awkwardly enough along the ground; but if brought into a wooded district, at once reverts to an arboreal life, thus showing how persistent are inherited instincts and habits. In describing a female and young of this variety, Mr. Hudson states that when the latter have attained the size of large Rats they are carried in all sorts of positions on the back of the mother, although from his figure they do not seem to twine their own tails round the parental tail in the manner characteristic of some of the other species. The mother being less than a Cat in size, the burden of carrying eleven young ones of the dimensions of Rats may well be imagined, yet even with such a tremendous load she is able to climb trees with activity and speed. If Mr. Hudson is right in his specific identification of the Opossum in question, it would appear that after a certain age the young of a pouched species may resort to the maternal back as a resting-place. From its Weasel-like shape, the Thick-tailed Opossum is admirably adapted to live in localities like the pampas, where the ground is covered for a large part of the year with luxuriant grass, between the stems of which the sharp nose of the creature is well suited to make its way.

V. PHILANDER OPOSSUM. DIDELPHYS PHILANDER.

Didelphys philander, Linn. Syst. Nat., ed. 10, vol. i., p. 54 (1760); Thomas, Cat. Marsup. Brit. Mus., p. 337 (1888).
Didelphys dicrura, et D. affinis, Wagner, Archiv. für. Nat., vol. viii., p. 358 (1842).

Characters.—With this species—the *Manicou gros eaux* of the French in South America—we come to the first of the two representatives of the sub-genus *Philander*. In addition to their woolly fur, and the presence of a dark stripe down the middle of the face, these two species present the following characteristics in common. Size medium; fourth hind toe the longest, the third and fifth next in size and about equal, and the second slightly the shortest of the four. Pouch rudimentary.

The Philander Opossum may be characterised as follows:—

Size generally smaller, form more slender, and tail relatively longer than in the preceding species; fur thick, soft, and woolly. General colour dull yellowish or rufous grey; face pale grey, with a distinct narrow median line, as well as the area round the eyes, brown; under-parts deep or pale yellow, without any line of demarcation from the colour of the back. Naked portion of muzzle slightly projecting backwards above, and with two distinct notches in the lip on each side of the median groove. Ears large and naked, with a well-developed anterior basal prolongation. Pouch represented only by lateral rudiments. Seven teats. Limbs dull grey, and the short and nearly naked, feet brown; foot-pads large, rounded, and slightly prominent, a minute extra one being generally present near the heel. Tail longer than the head and body, furred for two or three inches at the base, the limits of the fur stopping suddenly and forming a ring ; elsewhere naked, with almost imperceptible scales ; in colour, the base grey and the tip white, the two colours forming a series of mottlings at their junction. Length of head and body of male about 10 inches ; of tail 12½ inches. Females larger.

Distribution.—North-eastern South America, *e.g.*, Guiana and Brazil.

VI. WOOLLY OPOSSUM. DIDELPHYS LANIGERA.

Didelphys lanigera, Desmarest, Mamm., vol. i., p. 258 (1820) ; Thomas, Cat. Marsup. Brit. Mus., p. 339 (1888).
Didelphys derbiana, Waterhouse, Jardine's Nat. Library, Mamm., vol. xi., p. 97 (1841).
Didelphys cchropus, Wagner, Archiv. für Nat., vol. viii., p. 359 (1842).
Didelphys ornata, Tschudi, Fauna Peruana, Mamm., p. 146 (1844).

(*Plate XXXIV.*)

Characters.—Size rather larger than in the preceding species ; fur thick, soft, and even more woolly than in the latter. General colour varying from rich dark rufous to pale bright fawn, more or less variegated with white ; face greyish-white, with the dark stripe generally conspicuous, but ill-defined in some of the paler examples ; a reddish area round the eyes ; under-parts greyish-white, more or less tinged with rufous ; middle of front portion of back with a conspicuous greyish or greyish-white stripe, sometimes extending backwards to the tail, and in other cases absent ; sides of neck and back bright red ; limbs grey or pale rufous. Tail very long, furry at the basal end for from one-half to one-third of its length on the upper surface, while inferiorly the hair stops an inch or more short of the point which it reaches superiorly, the termination of the furry portion thus forming an oblique line on the sides ; naked portion grey basally and yellow terminally, with a mottling of

WOOLLY OPOSSUM

the two colours at their junction. Other characters apparently much the same as in the last species. Length of head and body of male about 12¼ inches; of tail about 15 inches.

Distribution.—South-eastern Mexico to Paraguay.

The specimen represented in Plate XXXIV. is the type of *Didelphys derbiana*, so named from having been originally in the collection of a former Earl of Derby at Knowsley. Mr. Waterhouse, its describer, believed it to represent a distinct species on account of its brilliant coloration, the well-marked white stripe on the withers, and the great relative length of the hairy portion of the tail. The soles of the hind feet are also peculiarly black, while the fore feet are furnished with white hairs, and the naked portion of the tail is clouded with brown. A larger series of specimens have, however, shown that such variations are merely individual and have no specific value.

This species is one of those in which the young, as soon as they leave the teats, are habitually carried on the back of the mother, with their prehensile tails tightly twisted round her tail. This feature, it may be added, is common to the whole of the small forms constituting the next sub-genus, one of which has a synonym (*D. dorsigera*) referring to it.

VII. ASHY OPOSSUM. DIDELPHYS CINEREA.

Didelphys cinerea, Temminck, Monogr. Mamm., vol. i., p. 46 (1827); Thomas, Cat. Marsup. Brit. Mus., p. 342 (1888).

Micoureus cinereus, Lesson, Nouv. Table Règne Anim., Mamm., p. 186 (1842).

Didelphys noctivaga, Tschudi, Fauna Peruana, Mamm., p. 148 (1844).

Didelphys waterhousei, Tomes, Proc. Zool. Soc., 1860, p. 58.

Characters.—With this species we reach the fourth sub-genus (*Micoureus*) of the true Opossums, which includes a considerable number of species, and is characterised as follows :—

Size small; form slender; tail long, and generally much exceeding the length of the head and body; relative lengths of the hind toes as in the preceding sub-genus, but in some species the fifth not longer than the second; pouch wanting; fur straight, slightly woolly in some species. The dark streak down the face characterising the two representatives of the preceding group is absent.

From the allied forms the species under consideration may be distinguished by the following characters:—

Size larger than in any other member of the group, although considerably inferior to that of the smallest of the species described above; fur soft, close, and slightly woolly; naked portion of muzzle as in *D. philander*, with two notches on each side of its lower margin. General colour clear grey, washed with yellowish on the sides, and not unfrequently tinged with rufous; face grey, with a more or less distinct black band through and round each eye; under-parts yellowish-white, with the hairs grey at the base; limbs grey; feet nearly naked, and either whitish or pale brown. Ears large and rounded, with a large pointed projection at the base. Soles of feet with large, rounded, and finely striated pads, six anterior and five or six posterior, that of the hallux not being fully divided. Nine teats. Tail furry for one or two inches at the base, then becoming naked and scaly, its colour slaty-grey basally and white or yellow terminally, without any mottling at the junction of the two colours. Length of head and body about 7 inches; of tail 9¾ inches.

Distribution.—Costa Rica to Brazil.

In size this species may be compared to the English Black Rat; but the form from Ecuador described as *D. waterhousei*, and identified with this species by Mr. Thomas, is smaller, and also differs in the length of the furred portion of the tail. It is noteworthy that in describing this form Tomes states that he had observed a female with young in her pouch, a statement which, if confirmed, would indicate not only the specific distinctness of *D. waterhousei*, but would likewise prove that it differed from all the other members of the sub-genus in the possession of a pouch.

Like the other members of the group, the Cinereous Opossum is essentially an arboreal, and mainly an insectivorous animal, climbing with the assistance of its prehensile tail, but otherwise very similar in its general habits to the Oriental Tree-Shrews.

VIII. MURINE OPOSSUM. DIDELPHYS MURINA.

Didelphys murina, Linn. Syst. Nat. ed. 10, vol. i., p. 55 (1760);
Thomas, Cat. Marsup. Brit. Mus., p. 343 (1888).
Didelphys dorsigera, Linn., *loc. cit.*
Micoureus murinus, et M. dorsigerus, Lesson, Nouv. Tabl. Règne Animal, Mamm., p. 186 (1842).
Didelphys impavida, Tschudi, Fauna Peruana, p. 149 (1844).
Didelphys musculus, Cabanis, in Schomburgk's Reis. Guian., vol. iii., p. 778 (1848).

Characters.—Size considerably less than that of the preceding species; form slender and delicate; fur thick, close, and straight, without any tendency to woolliness; muzzle with two inferior notches on each size. General colour deep dull rufous, varying considerably in intensity and shade; face greyer and paler than the back, with the dark eye-stripes generally very conspicuous, and contrasting sharply with the pale ground-colour; chin white; rest of under-parts yellowish white; outer sides of limbs like back, inner surface white; feet white or pale grey. Ears very similar to those of the Ashy Opossum. Teats from nine to fifteen in number. Foot-pads as in preceding species. Tail long, slender, cylindrical; furry for about half-an-inch at the base, elsewhere with only a few scattered hairs;

colour grey, sometimes becoming lighter near the tip. Length of head and body of male about 5½ inches; of tail 8½ inches; female slightly larger, with a shorter tail.

Distribution.—Central Mexico to Brazil.

Habits.—As the preceding species was compared in size to the Black Rat, so the present one, which is the best known representative of the group, may be likened to a common Mouse in this respect. Very abundant in Guiana, it is said not only to be as expert a climber as its kindred, but also at times to burrow in the ground. In addition to insects, it preys upon small birds, and is reported not to disdain fruits.

IX. TEHUANTEPEC OPOSSUM. DIDELPHYS CANESCENS.

Didelphys (Micoureus) canescens, J. A. Allen, Bull. Amer. Mus. Nat. Hist., vol. v., p. 235 (1893).

Characters.—Size even smaller than that of *D. murina*; fur short, thick, and close. General colour (very similar to that of the undermentioned *D. grisea*) ashy brown, with a slight rufescent tinge in some specimens; under-parts white, tinged with pale yellow; a broad blackish eye-ring, extending forwards nearly to the nose; area between the eyes, sides of face, and neck yellowish-grey, and much lighter than the back. Ears broad, rounded, naked, probably yellowish in life. Tail slightly longer than the head and body, heavily furred for the basal half-inch, elsewhere naked; in colour pale brown, either uniform, or variegated with flesh-coloured spots. Length of head and body about 5¼ inches; of tail 5¾ inches.

Distribution.—Isthmus of Tehuantepec, Mexico. Compared with *D. murina*, which it much resembles in colour, this species is fully one-third smaller, with a relatively shorter tail in which the furred portion is of less extent. Possibly it may prove identical with *D. waterhousei*, if that be distinct from *D. murina*;

if so, the former will agree with the other members of the subgenus in being devoid of a pouch.

X. SHINING OPOSSUM. DIDELPHYS LEPIDA.

Didelphys (*Micoureus*) *lepida*, Thomas, Ann. Mag. Nat. Hist., ser. 6, vol. i., p. 158, and Cat. Marsup. Brit. Mus., p. 347 (1888).

Characters.—Size very small; fur soft, close, and straight. General colour full rich rufous, of a much deeper tone than in any of the allied species; under-parts dirty white, with a faint tinge of rufous.

The species may be readily distinguished from *D. murina* by its smaller and shorter ears, from the undermentioned *D. pusilla* by the presence of projections at the anterior margin of the ears, and by ledges on the skull above the sockets of the eyes; while it differs from both in its more brilliant coloration and the smaller number of teats (seven).

Distribution. - Amazonia, from the Peruvian Andes, to Guiana.

XI. PIGMY OPOSSUM. DIDELPHYS PUSILLA.

Didelphys pusilla, Desmarest, Nouv. Dict. d'Hist Nat., vol. xxiv., Table, p. 19 (1804); Thomas, Cat. Marsup. Brit. Mus., p. 348 (1888).

Didelphys nana, Illiger, Abhandl. Akad. Berlin, 1811, p. 107 (1815).

Didelphys (*Grymæomys*) *agilis*, Burmeister, Thiere Brasil., vol. i., p. 139 (1854).

Micoureus pusillus, Gervais, in Castelnau's Voy. Amer. Sud. Mamm., p. 103 (1855).

Characters.—Size nearly as in the last; fur soft, thick, and straight; naked portion of muzzle with two distinct notches on each side of its lower border. General colour bright rufous, somewhat paler than in the last species; the uniform rufous of the back gradually becoming paler on the flanks, till it passes into whitish-rufous or white on the under surface; black eye-mark very distinct, and extending forwards on the sides of the muzzle nearly to the nose; inner sides of limbs like under-parts; feet white or pale brown, nearly naked. Ears large, with the anterior basal projection, short, rounded, and frequently almost wanting. Thirteen or fifteen teats. Fifth hind toe very slightly shorter than the second; foot-pads, low, rounded, and but slightly prominent. Tail long, slender, tapering, with its base scarcely furred, in colour uniform pale-grey, becoming lighter inferiorly. Length of head and body in male about 3¼ inches; of tail 4 inches; females slightly larger.

Distribution.—Brazil, from Santarem to Rio Grande do Sul.

Nothing specially noteworthy appears to have been recorded as to the habits of this beautiful little creature.

XII. GREY OPOSSUM. DIDELPHYS GRISEA.

Didelphys grisea, Desmarest, Dict. Sci. Nat., vol. xlvii., p. 393 (1827); Thomas, Cat. Marsup. Brit. Mus., p. 349 (1888).

Didelphys incana, Lund, Blik. Brasil. Dyr., Dansk. Afhand., vol. viii., p. 237 (1841).

Micoureus griseus, Gervais, Hist. Nat. Mamm., vol. ii., p. 287 (1855).

Characters.—Size nearly as in the Murine Opossum; fur close, soft, and somewhat fluffy. General colour uniform deep grey, with scarcely a tinge of rufous; face somewhat paler, with the dark eye-mark inconspicuous and confined to the front of the eyes; under-parts pure white, sharply defined from the grey of the back, the line of demarcation being often indicated by a tinge of fulvous or rufous; back of fore limbs white, sometimes a white ring encircling the elbow; back of fore feet pale brown, of hind feet white. Ears very large and leaf-like, with the anterior basal projection small and rounded. Toes and foot-pads as in the Murine Opossum. Number of teats unknown. Tail long, slender, and tapering, with the basal half-inch furry, elsewhere naked; in colour grey above and whiter inferiorly. Length of head and body of female about 5 inches; of tail 7 inches.

Distribution.—Central and Eastern Brazil.

XIII. VELVETY OPOSSUM. DIDELPHYS VELUTINA.

Didelphys velutina, Wagner, Archiv. für. Nat., vol. viii., p. 360 (1842); Thomas, Cat. Marsup. Brit. Mus., p. 352 (1888).
Microdelphys velutina, Burmeister, Erläut. Faun. Brasil., p. 86 (1856).

Characters.—Size small; fur peculiarly soft, crisp, and velvety. General colour soft Mouse-grey, suffused with dull rusty-brown along the sides; face paler, with the eye-mark indistinct; chin yellowish-white; chest rusty-fawn; remainder of under-parts pale cream-colour, with the bases of the hairs dark slaty, the line of demarcation between the colours being well-defined; inner sides of limbs and the feet coloured like the under-parts. Ears large, with a slight convex anterior basal projection. Fifth hind toe reaching to the end of the second joint of the fourth. Tail shorter than the head and body, with its basal half-inch thickly furred. Length of head and body of male about $3\frac{1}{3}$ inches; of tail $\frac{4}{5}$ inches.

Distribution,—S. Paulo, Brazil.

This very rare form appears separable from the Grey Opossum solely on account of the shortness of the tail; and additional specimens are needed to show whether this feature is constant.

XIV. CHILIAN OPOSSUM. DIDELPHYS ELEGANS.

Didelphys elegans, Waterhouse, Voy. H.M.S. "Beagle," Mamm., p. 95 (1839); Thomas, Cat. Marsup. Brit. Mus., p. 353 (1888).
Micoureus elegans, Lesson, Nouv. Tabl. Règne Animal, Mamm., p. 186 (1842).

Characters.—Size nearly the same as in the Grey Opossum, or rather smaller; fur long, soft, and silky. General colour soft pale grey, finely grizzled with reddish-brown; middle of face pale grey, with the eye-markings forming a distinct ring round each eye, and extending only a short distance forwards on the sides of the muzzle; under-parts, at least in the middle line, pure white, with the line of demarcation not very well defined; legs white internally; feet pale brown. Muzzle long and pointed, with only a single inferior notch in the naked portion on each side of the median groove. Ears very large, narrow, and oval, with the anterior basal projection almost or quite absent. Front and hind feet with six tall and prominent pads, between which the sole is granulated. Tail with the basal half-inch thin, the next two inches much thickened, and the terminal half rapidly tapering to a point; in colour grey, with minute white hairs scattered over it. Length of head and body about 3¾ inches; of tail 4¾ inches.

Distribution.—South Brazil and Chili.

Habits.—This Opossum, which was first brought to the notice of scientists by Darwin, who obtained it at Valparaiso during the voyage of H.M.S. "Beagle," may be easily recognised by the peculiar thickening of the tail at a short distance below the root. It appears to be very abundant in Chili, where it extends as far north as Cobija. According to Darwin, it is found in the thickets clothing the rocky hills near Valparaiso. There, he writes, these little creatures "are exceedingly numerous, and are easily caught in traps, baited with either cheese or meat. The tail appeared to be scarcely at all used as a prehensile organ; they are able to run up trees with some degree of facility. I could distinguish in their stomachs the larvæ of beetles."

XV. YELLOW-FLANKED OPOSSUM. DIDELPHYS DIMIDIATA.

Didelphys dimidiata, Wagner, Abhandl. Akad. München, vol. v. p. 151 (1847); Thomas, Cat. Marsup. Brit. Mus. p. 355 (1888).

Microdelphys brachyura, Burmeister, Erläut. Faun. Brasil., p. 86 (1856).

Characters.—The present, together with the nine remaining representatives of the True Opossums, constitute the sub-genus *Peramys*, characterised as follows :—

Size small; tail short, generally equal to about half the length of the head and body, more or less covered with short, fine hairs, and but slightly, if at all, prehensile; fifth hind toe considerably shorter than the second; third and fourth subequal, and only slightly exceeding the second in length.

Were it not that these small species are approached by the Velvety Opossum in the shortness of the tail, they might, writes Mr. Thomas, be regarded as representing a distinct genus, in spite of the circumstance that they show no absolutely distinctive peculiarities either in the skull or the teeth. Although their habits have never been properly described, it appears—as may be inferred from the lack of prehensile power in the tail—that they are far less arboreal than the other members of the genus. On this account, as already mentioned, they may be regarded as occupying in South America the place in nature elsewhere filled by the Shrews.

The present species, which was described by Waterhouse under the name of *D. brachyura*, a name previously used by

Schreber for the next member of the genus, may be characterised as follows :—

Size large; fur short, coarse, and harsh. General colour pale grizzled grey on the crown of the head and middle of the back, and rich orange yellow on the flanks and under-parts; legs yellow; feet greyer. Muzzle long and slender, with the naked portion extending backwards as a short prolongation in the middle line. Ears rounded and very short, with the anterior basal projection well-developed. A gland on the chest in the male. Skin of soles of feet rough and coarse; pads of hind feet only five in number, and not so well-defined as in the other members of the group. Tail about half the length of the head and body, thick, regularly tapering, and clothed with numerous short fine hairs; in colour brown above, and yellowish beneath. Length of head and body in a large male about 6 inches; of tail $3\frac{1}{2}$ inches.

Distribution.—South Brazil and Uruguay.

XVI. RED-FLANKED OPOSSUM. DIDELPHYS BREVICAUDATA.

Didelphys brevicaudata, Erxleben, Syst. Regn. Anim., p. 80 (1777); Thomas, Cat. Marsup. Brit. Mus., p. 356 (1888).
Didelphys brachyura, Schreber, Säugethiere, vol. iii., p. 548 (1778).
Didelphys tricolor, Geoffr., Cat. Mus., p. 144 (1803).
Didelphys hunteri, Waterhouse, Jardine's Naturalist's Library, Mamm., vol. xi., p. 110 (1841).
Peramys tricolor, et P. brachyurus, Lesson, Nouv. Tabl. Règne Animal, Mamm., p. 186 (1842).
Didelphys glirina, Wagner, Archiv. für Nat., vol. viii., p. 359 (1842).

Characters.—May be distinguished from the preceding by the large ears and the red colour of the flanks; the number of teats (unknown in the yellow-flanked species) being either five or seven, one of which is placed as a centre and surrounded by the others.

Distribution.—Guiana and Brazil.

XVII. GREY-FACED OPOSSUM. DIDELPHYS DOMESTICA.

Didelphys domestica, Wagner, Archiv. für. Nat., vol. viii., p. 359 (1842); Thomas, Cat. Marsup. Brit. Mus., p. 358 (1888).

Characters.—Size large; fur thick, straight, and soft. General colour pale grey, without dark eye-marks; under-parts white or greyish-white, sometimes tinged with yellow. Ears large and rounded. Teats thirteen, of which three are central, and the remainder arranged in lateral pairs. Fifth hind toe reaching to the middle of the second joint of the fourth. Tail little more than half the length of the head and body, its basal half-inch furred and grey, elsewhere short-haired and dark brown. Length of head and body of male about 5¼ inches; of tail 2¾ inches.

Distribution.—Brazil.

XVIII. RED-FACED OPOSSUM. DIDELPHYS SCALOPS.

Didelphys (Peramys) scalops, Thomas, Ann. Mag. Nat. Hist., ser. 6, vol. i., p. 158, and Cat. Marsup. Brit. Mus., p. 359 (1888).

Characters.—Form rather more light and slender than in the last species, from which this one may be distinguished by the rufous head, rump, and tail, and the grey fore part of the back and under-parts. The number of teats is unknown.

Distribution.—Brazil.

This Opossum, in which the distribution of the rufous and grey forming its general colour is very remarkable, was described on the evidence of two specimens in the British

Museum, but others are required in order to determine whether the pattern of colour is constant.

XIX. HENSEL'S OPOSSUM. DIDELPHYS HENSELI.

Didelphys henseli, Thomas, Ann. Mag. Nat. Hist., ser. 6, vol. i., p. 159, and Cat. Marsup. Brit. Mus., p. 360 (1888).

Characters.—While agreeing with the preceding members of the sub-genus in the absence of dark stripes on the back, this short-tailed Opossum is distinguished by its inferior size, and the presence of about twenty-five teats, of which five are central and the remainder arranged in lateral pairs. The ears are small; while in colour the back is grey and the flanks are red. The tail is about equal in length to the head and body, with its root alone furry, and the remainder scaly and nearly naked, its upper surface being brown, and the under side red. Length of head and body of male about $4\frac{1}{4}$ inches; of tail $2\frac{1}{5}$ inches.

Distribution.—Entre Rios and Rio Grande do Sul. From the number of teats it is probable that the female of this species gives birth to a large progeny, and it would be interesting to ascertain in what manner the young are carried about by such a diminutive creature.

XX. SHREW-OPOSSUM. DIDELPHYS SOREX.

Microdelphys sorex, Hensel, Abhandl. Akad. Berlin, 1872, p. 122.

Didelphys sorex, Thomas, Cat. Marsup. Brit. Mus., p. 362 (1888).

Characters.—This species may be easily recognised by being the smallest representative of the sub-genus with a uniformly-coloured back. The ears are of medium size; and the general colour grey above and rufous on the flanks, the latter colour extending on to the cheeks and hips. Number of teats un-

PLATE XXXV

known. Length of head and body of male about 2¾ inches; of tail 1¾ inch.

Distribution.—Rio Grande do Sul.

XXI.˙ THREE-STRIPED OPOSSUM. DIDELPHYS AMERICANA.

Sorex americanus, P.L.S., Müller, Lin. Natursyst. Suppl., vol. vii., p. 36 (1776).
Sorex braziliensis, Erxleben, Syst. Regn. Anim., vol. i., p. 127 (1777).
Didelphys tristriata, Illiger, Abhandl. Akad. Berlin, 1811, p. 107 (1815).
Didelphys trilineata, Lund, Blik. Brasil. Dyrev., Dansk. Afhand., vol. viii., p. 237 (1841).
Peramys tristriata, Lesson, Nouv. Tabl. Règne Anim., Mamm., p. 187 (1842).
Didelphys americana, Thomas, Cat Marsup. Brit. Mus., p. 363 (1888).

(*Plate XXXV.*)

Characters.—This medium-sized and Shrew-like species is sufficiently distinguished from all the foregoing short-tailed Opossums by the presence of three dark longitudinal lines running down the back; the general ground-colour being grey or rufous. There are fifteen pairs of teats, of which five are central. Length of head and body of male about $5\frac{1}{2}$ inches; of tail $2\frac{2}{5}$ inches; female rather smaller, with the tail relatively longer.

Distribution.—Brazil.

Till the publication of the British Museum Catalogue of Marsupials in 1888, this species was almost universally known by the appropriate name of *Didelphys tristriata*, and since the object of nomenclature is to enable us to recognise and distinguish animals with facility, it seems a pity that a rigid adherence to the rule of priority should have led to the substitution of such an utterly meaningless name as *Didelphys americana*. Still, it must be confessed that if we once break through the rule of adopting the earliest specific name proposed for an animal as its proper designation, it becomes very difficult to know where to stop.

XXII. LESSER THREE-STRIPED OPOSSUM. DIDELPHYS IHERINGI.

Didelphys (Peramys) iheringi, Thomas, Ann. Mag. Nat. Hist., ser. 6, vol. i., p. 159, and Cat. Marsup. Brit. Mus., p. 364 (1888).

Characters.—Although this tiny little species appears at first sight to be nothing more than a dwarf race of the preceding, it may, according to its describer, be distinguished not only by its inferior size, but likewise by the different conformation of the skull, and more especially by the marked flattening of the region of the forehead. Length of head and body of male about 3 inches; of tail 1¾ inch.

Distribution.—South Brazil.

XXIII. SINGLE-STRIPED OPOSSUM. DIDELPHYS UNISTRIATA.

Didelphys unistriata, Wagner, Archiv. für Nat., vol. viii., p. 360 (1842); Thomas, Cat. Marsup. Brit. Mus., p. 365 (1888).

Characters.—Apparently only known by a single specimen, this litttle short-tailed Opossum is sufficiently distinguished by the single dark reddish-brown line running down the middle of the back from behind the shoulders to the rump; the general colour of the upper surface being pale grizzled grey, and the hairs tipped with rufous, while the flanks and under-parts are bright orange, as are the greater portion of the limbs. The ears are very short, rim-like, and nearly naked; and the tail has its basal half-inch furred, and gradually passing into the

WATER OPOSSUM.

short-haired remainder. Length of head and body of male about 5½ inches; of tail 2½ inches.

Distribution.—S. Paulo, Brazil.

NOTE.—Mr. Oldfield Thomas informs me, on the authority of Dr. Goeldi, that the Opossum called by Burmeister *Microdelphys alboguttata* (Th. Bras. i., p. 340) is nothing but a young specimen of the common *Dasyurus viverrinus* of Australia, the Brazilian locality being, of course, erroneous.

THE WATER-OPOSSUMS. GENUS CHIRONECTES.

Chironectes, Illiger, Prodromus Syst. Mamm., p. 76 (1811).

Distinguished from *Didelphys* by the presence of a prominent tubercle on the inner side of the fore foot simulating a sixth toe, and by the hind feet being webbed as far as the ends of the toes, so that only their terminal pads project beyond the webbing. Owing to the first hind toe, or hallux, being included in the webbing, it is much less opposable than in the True Opossums.

According to the author of the British Museum Catalogue of Marsupials, the single representative of this genus is more nearly allied to the True Opossums of the sub-genus *Metachirus* than to any of the others.

1. WATER-OPOSSUM OR YAPOCK. CHIRONECTES MINIMUS.

Lutra minima, Zimmermann, Geograph. Geschicte, vol. ii., p. 317 (1780).

Didelphis minima, Cuvier, Tabl. Elem., p. 125 (1798).

Chironectes minimus, Illiger, Prodromus Syst. Mamm., p. 76 (1811); Thomas, Cat. Marsup. Brit. Mus., p. 368 (1888).

Chironectes variegatus, Illiger, Abhandl. Akad. Berlin for 1811, p. 107 (1815).

Chironectes palmata, Cuvier, Règne Animal, vol. i., p. 174 (1817).

Chironectes yapock, Desmarest, Mammalogie, vol. i., p. 261 (1820).

(*Plate XXXVI.*)

Characters.—Size about equal to that of the common Opossum; fur very thick, close, and woolly, having a few longer straight hairs intermingled with the shorter ones; naked portion of muzzle with a short backward extension in the middle line; ears large and rounded, with the anterior basal projection rudimental. Face, in addition to the usual "whiskers" on the sides of the muzzle, with a series of tufts of long and stout bristles, one pair of which is situated above the eyes, another in front of and below the eyes, and a single median one on the throat. General colour greyish-white, marbled with dark brown; the muzzle, and a streak running through each eye to below the ear, as well as the top of the crown of the head, deep blackish-brown, a conspicuous greyish-white crescentoid band passing above the eyes to the bases of the ears; on the back a black stripe running from the head to the root of the tail, and spreading out on the sides into four well-defined patches, of which one is situated on the shoulder, another in the middle of the back, the third on the loins, and the fourth on the rump, the ground-colour between them being slaty grey; under-parts and inner sides of the legs white; feet brown and nearly naked, with the toes silvery grey. Soles of feet with a uniform fine granulation, the pads being almost wanting. Tail long and powerful, furred for the basal one or two inches, and the fur extending further on the upper and lower surfaces than on the sides; elsewhere naked and coarsely scaled, the basal portion being black, and the extremity yellowish. Length of head and body about 14 inches; of tail $15\frac{1}{2}$ inches.

Distribution.—Guatemala to South Brazil.

Habits.—The Yapock, or Water-Opossum, which may be compared in size, as also to some extent in form and proportions, to a large Rat, is a comparatively rare animal inhabiting the rivers of Brazil, and living chiefly or exclusively on crustaceans and insects. Although very little is known of its habits, it would appear from the circumstance that a specimen was taken in a basket-trap similar to those used for catching eels in Europe, that the creature is an expert diver. Small fish doubtless also form a portion of its diet. The female has a well-developed pouch, in which the young, usually five in number, are carried for some time; and it will be obvious that during that period the creature must refrain from entering the water. Later on, the young accompany their parent to the river, and are exercised by her in swimming and diving.

PART II.

MONOTREMES—ORDER MONOTREMATA.

INTRODUCTION.

AT the commencement of the first part of this work it was shown that the Marsupials, or Pouched Mammals, constitute not only a distinct order (Marsupialia) in the Mammalian class, but that they likewise form a separate sub-class variously known as the Didelphia, Metatheria, or Implacentalia, in opposition to the group known as the Monodelphia, Eutheria, or Placentalia, which contains all the other members of the class save those to be now considered.

The representatives of this latter group, which are very few in number, and strictly confined to Australia, Tasmania, and New Guinea, are known as Monotremes, or Egg-laying Mammals, and agree with the Marsupials in constituting not only a distinct order (Monotremata), but likewise a separate sub-class; —the sub-class being designated by the name of Ornithodelphia, or Prototheria. So vast indeed is the difference between these Egg-laying Mammals and all other members of the class to which they belong, and so closely do they connect the latter with the lower classes of the Vertebrate sub-kingdom, that some systematic zoologists do not consider that their separation as a sub-class of equivalent rank with those respectively containing the Pouched and the Placental Mammals, sufficiently emphasizes their radical distinction from both the latter groups. Accordingly it has been proposed to brigade both the Pouched and the Placental Mammals in one grand division of the class, and to make the Egg-laying Mammals the sole representatives of a second division of similar rank. Although there is a good deal to be said in favour of such a binary division of the Mammalian class, yet as the ternary division has been almost universally adopted, it seems a pity to disturb it, more especially as it serves all the purposes of classification. In following such a ternary arrangement, all that the student has to bear in mind is that the gap separating the third sub-class from the second is vastly greater than that by which the first and second are sundered.

With these general preliminary remarks, we may proceed to the consideration of the characters by which the aforesaid Monotremes or Egg-laying Mammals are thus sharply distinguished from all other living representatives of the class. It may be mentioned, however, that it is only comparatively recently that it has been definitely ascertained that these strange Mammals, although nourishing their young with milk secreted from special glands on the body of the female parent, after the manner of the rest of their class, yet actually lay eggs like a Bird or a Reptile. Reports were, indeed, brought home by those who had interviewed the aborigines of Australia as to the oviparous habits of one at least of these animals, but these were generally discredited either by the interviewers themselves, or by those to whom the story was reported. Still, however, even among anatomists and zoologists themselves there seems to have been a lingering suspicion that there was something peculiar about the reproduction of these antipodean creatures. For instance, Meckel the anatomist, who first proved the existence in one of them (the Duck-bill) of mammary or milk-glands, still doubted whether the animal might not lay eggs after the manner of Birds; this opinion being based on the similarity between the reproductive organs of the Monotremes and those of Birds. After various other naturalists had taken part in the discussion, some of whom even denied the existence of milk-glands, or, if they admitted these, regarded them as different from those of other Mammals, Owen entered the arena, and, after fully confirming Meckel's demonstration of the existence of the glands in question, decided that the Monotremes hatched their eggs within their own bodies and brought forth living young, which were suckled by their female parent until able to shift for themselves.

The great authority of Owen on all questions connected with anatomy and zoology seems to have led to the oo-viviparous reproduction of the Monotremes being accepted as a settled fact for many years. This was confirmed in a work published in 1860, by Dr. George Bennett, who devoted much time and labour to the endeavour to elucidate the breeding habits of one of these creatures, and who, in spite of the somewhat confused accounts of the natives to the contrary, came to the conclusion that in all probability these animals did not lay eggs. Thus matters stood for many years, and it was not until 1884 that it was conclusively proved that the Monotremes did actually lay eggs similar in structure to those of Birds and Reptiles.

Curiously enough, there was another deficiency in our knowledge of one of these animals—the Duck-bill,—which was not remedied till 1888, and then only partially so. This relates to the alleged absence in this creature of teeth, the place of which was believed to be supplied by a series of horny plates on the palate. Nevertheless, it was finally proved that during a certain portion of their existence these animals do possess well-developed teeth; such teeth being of a very peculiar and remarkable type.

The Monotremes derive their name from the circumstance

that there is, as in Birds and Reptiles, but a single aperture at the hinder extremity of the body from which are discharged the whole of its waste-products, together with the reproductive elements; the oviducts opening separately into the extremity of this passage, which is termed the cloaca. Reproduction is effected by means of eggs, which are laid and hatched by the female parent; while, after the extrusion from the egg, the young are nourished by milk secreted by special glands situated within a temporary pouch, into which the head of the young animal is inserted and retained. The ducts from these milk-glands instead of opening into teats, discharge merely upon the porous skin of the inside of the pouch by means of numerous apertures; the milk being forced into the mouths of the young by the contraction of special muscles. The milk-glands are of simpler structure than those of other Mammals, and, according to Gegenbaur's researches, correspond to the sweat-glands of the latter, and not to the milk-glands; so that these milk-glands of the Monotremes do not correspond homologically with those of other members of the class.

The skeleton differs from that of the higher Mammals, and thereby corresponds with that of certain lower Vertebrates, in regard to the structure of the shoulder-girdle, or that portion which serves for the support and attachment of the bones of the fore limb. In nearly all the higher Mammals the shoulder-girdle comprises merely a pair of collar-bones, or clavicles, and two blade-bones, or scapulæ; each of the latter having a projection on its lower extremity termed the coracoid process, —so-called on account of a fancied resemblance in man to the beak of a Raven. In the Monotremes, on the other hand, there is an additional T-shaped unpaired bone lying on the collar-bones and breast-bone, known as the interclavicle; this bone occurs in no other Mammals, but is well-developed in Lizards and many extinct Reptiles. In place of the coracoid process of the blade-bone, there are on each side two distinct bones lying on the inferior aspect of the skeleton, and connected with the former. Of these, the most anterior, as representing the coracoid process of the blade-bone, may be termed the coracoid; while the posterior element may be called the metacoracoid. Both these pairs of bones exist in certain extinct Reptiles; while the hinder pair are represented in Birds, where they are commonly termed the coracoids. The metacoracoids, it should be added, are articulated with the breast-bone, or sternum, as in Birds.

The brain, although of a distinctly Mammalian type, is very simple; and has the connecting cross-fibres on the lower surface known as the corpus callosum, either very slightly developed, or not present at all; Mr. A. Hill, who is one of the latest investigators on the subject, believing in the complete absence of the structure in question. The minute bones occurring in the interior of the ear are likewise of a much simpler type than in other Mammals; and they once more show the affinities of these creatures to the lower classes of Vertebrates.

The above are some of the most important features which justify the separation of the Monotremes not only as an Order, but likewise as a distinct Sub-Class; and they serve to show how very widely the group differs from all other Mammals. It is true that they agree with the latter in the circumstance that the young are nourished by means of milk-secreting glands; but if Gegenbaur be right in his view that these glands in the Monotremes are not homologous with those of other Mammals, it follows that the milk-secreting function is of no classificatory value whatever; the presence of these glands in the Monotremes on the one hand and in the higher Mammals on the other being an instance of that parallelism in development to which allusion has already been made. If, therefore, any modification in the usually accepted classification is advisable, there would seem to be considerable justification for the removal of the Monotremes from among the Mammals to form a separate class by themselves; although such a radical change should only be made after much deliberation and careful weighing of evidence.

As characters which may be regarded as distinctive of the Monotremes as an Order, in contradistinction to a Sub-Class, the following may be noted :—

In both the two existing families the males are provided with a perforated spur on the inner side of the heel, connected with a gland behind the thigh, the function of which is not yet clearly understood. Although the assumption that this gland is poison-secreting has not been proved to be true, it has been suggested that during the breeding season if such a function were developed it might be advantageous to the males in their conflicts for the possession of the females.

Like the Marsupials, the Monotremes possess the so-called marsupial, or epipubic bones attached to the anterior rim of the pelvis. In all the muzzle is produced into a beak, which may be either flattened or cylindrical; and true functional teeth are absent at least in the adult. The external aperture of the ear opens on the surface of the head without any trace of a projecting conch; the short and powerful limbs are of subequal length, and adapted for digging; and the tail is either short and broad, or rudimental.

That the existing Monotremes are specialised survivors of an extremely ancient stock, is rendered evident on the one hand by their peculiar beaks and the absence of teeth in the adult or throughout life, and on the other by their low general organisation and their marked indications of affinity with the lower Vertebrates. On the whole, their nearest allies among the latter seem to be a group of Reptiles whose remains are found abundantly in the Secondary rocks of South Africa, and which are known as Anomodonts; the Reptiles in question being the only known Vertebrates having a shoulder-girdle of the type characterising the Monotremes. Assuming these Anomodonts to be more or less nearly related to the ancestors of the latter, we are, however, still quite in the dark as to the intermediate links; as we are also in regard to the genetic relationship of the Monotremes to what we may term the True Mammals. There occur, however, in the Secondary and Tertiary rocks of many parts of the world certain very remarkable extinct Mammals which from the peculiar characters of their molar teeth are termed Multituberculata. Now these teeth present a certain distant resemblance to those of the Duck-bill among the Monotremes, and this, together with certain other evidence, may be indicative of some relationship between the two groups. The relationship must, however, at best, be but distant, and since the majority even of these ancient Multituberculates appear to have been comparatively specialised animals, it will be evident that we are still a very long way off the discovery of the true phylogeny of the Monotremes.

It may be added that in the Duck-bill the temperature of the blood is some 20° F. lower than in ordinary Mammals.

THE DUCK-BILLS. FAMILY ORNITHORHYNCHIDÆ.

Aquatic fluviatile Monotremes, in which the sexes differ markedly in size, presenting the following characters:

Muzzle in the shape of a broad, flattened beak, covered with delicate and sensitive skin, assuming in the dried state a horny consistence; tongue not extensile; fur without spines; tail well-developed, broad and flattened; feet modified into swimming organs, with the toes broadly webbed, and the soles

PLATE XXXVII

DUCK BILL

naked and devoid of pads. Three pairs of many-cusped molar-teeth in each jaw, of which the foremost is the smallest in the upper, and the hindmost in the lower jaw; these teeth disappearing in the adolescent animal, and their place taken in the adult by hard horny plates. Horny spur on hind foot of male very large. Hemispheres of the brain smooth.

The single genus and species by which the family is now represented is confined to Australia and Tasmania.

THE DUCK-BILLS. GENUS ORNITHORHYNCHUS.

Ornithorhynchus, Blumenbach, Voigt's Mag. Naturkunde, vol. ii., p. 205 (1800).

Form elongate and depressed; each foot with five toes, severally furnished with long claws, those on the fore feet being broad and blunt, while those of the hinder pair are compressed and pointed. Beak smooth, short, and evenly rounded, its junction with the head, both above and below, defined by a flap of skin; cheeks furnished with pouches.

THE DUCK-BILL. ORNITHORHYNCHUS ANATINUS.

Platypus anatinus, Shaw, Natur. Miscell., vol. x., pl. 385 (1799).
Ornithorhynchus paradoxus, Blumenbach, Voigt's Mag. Naturkunde, vol. ii., p. 205 (1800).
Ornithorhynchus rufus, et O. fuscus, Péron and Lesueur, Voyage Terres, Austral., atlas, pl. xxxiv. (1807).
Ornithorhynchus brevirostris, Ogilby, Proc. Zool. Soc., 1831, p. 150.
Ornithorhynchus crispus, et O. lævis, Macgillivray, Mem. Wern. Soc., vol. vi., pp. 128, 132 (1832).
Ornithorhynchus anatinus, Gray, List Mamm. Brit. Mus., p. 191 (1843); Thomas, Cat. Marsup. and Monotr. Brit. Mus., p. 388 (1888).

(*Plate XXXVII.*)

Characters.—Male markedly larger than female; fur of two kinds, the longer crisp, shining, and sometimes curly, and the under-fur short, soft, and woolly. General colour deep umber, or blackish-brown, tending in some specimens more to red and in others to black; covering of beak black above, and yellow and black beneath; a white or yellowish ring surrounding each eye; under-parts dirty greyish-white; with the hairs grey at the base and white at the tip; not unfrequently patches of dull chestnut on the chin, round the insertions of the limbs, and along the middle of the hinder portion of the under surface of the body. Tail coloured above like the back, generally naked inferiorly in the adult, but if hairy, whiter than the under-parts. Length of head and body of male about 18 inches; of tail 6 inches; the head and body in the female being about 4, and the tail 2 inches shorter.

It will be generally found in descriptions of the Duck-bill, Duck-Mole, or Water-Mole, as the animal is indifferently called, that the covering of the beak is stated to be of a horny or leathery nature. According to recent observations, however, this is not the case in the living animal, in which the muzzle is covered with a soft skin, comparable to that investing the nose of a Dog, and richly supplied with tactile nerves; such a structure being admirably adapted for the needs of a creature which is in the habit of raking and grovelling in the soft mud at the bottom of rivers in search of the small molluscs, insects, larvæ, and worms, which constitute its food. The snout is supported by a cartilage, which is believed to be a remnant of the primitive one upon which the bones were subsequently developed.

In the adult Duck-bill, in which the extremities of the muzzle are curiously expanded and flattened in a spatulate-like form, there are two pairs of horny plates, or cornules, of which the front ones are the most prominent and specialised. These anterior cornules are long and narrow, each forming a single longitudinal ridge; while each of the hinder pair consists of a broad horny and cuspidate plate, subdivided by transverse ridges into three cavities of different sizes. These cornules are developed from the mucous membrane of the mouth under and around the teeth, and they have beneath them hollows in the bones which are the remnants of the sockets of the latter. The teeth, whose function is subsequently performed by these horny plates, are broad and shallow, and after being gradually worn away by the sand swallowed with the food and the process of mastication, are finally shed ere the animal attains its full maturity. Whether these teeth correspond to the milk— or the permanent dentition of higher Mammals, has not yet been ascertained; but it is not improbable that they may represent the former. The teeth themselves are broad, flat, and low-crowned; the upper ones (with the exception of the minute anterior pair) having two tall cusps on the inner side, from which small ridges run downwards and outwards to the outer side; while the latter has a peculiarly crenulated edge. As already stated, these teeth are unlike those of any other existing Mammal, although they present a very distant approximation to those of the Banded Ant-eater. Their somewhat less remote resemblance to those of the extinct Multituberculate Mammals has been already mentioned.

Distribution.—Queensland to the south of latitude 18°, New South Wales, Victoria, South Australia, and Tasmania.

History and Habits.—The Duck-bill was originally described under the name of *Platypus anatinus*, which was Anglicised into Duck-billed Platypus, but since the generic name had been previously employed for another group of animals, it had, by the rules of zoological nomenclature, to give place to the later *Ornithorhynchus*, although Shaw's specific name of *anatinus* still holds good. On these grounds it is likewise preferable to discard the Anglicised term Duck-billed Platypus in favour of the simpler Duck-bill or Duck-Mole.

When the first stuffed example of this extraordinary animal reached Europe, it was boldly asserted that the specimen was a fraud, made up of a Duck's bill affixed to the skin of some large Mole-like creature. When, however, other examples were received, and it became certain that the Duck-bill was a genuine animal, the most unbounded curiosity and interest were aroused by such a strange creature. Curiously enough, it was then regarded as a connecting link between Mammals and the lower Vertebrates; the main ground for this view being the Duck-like beak, which, of course, is a mere specialised peculiarity of the animal, and has nothing whatever to do with its many bird-like and reptilian features. The long period that elapsed before the Duck-bill was definitely proved to lay eggs, and the still longer time that elapsed between the date of its first description and the discovery that in the earlier stages of its existence the creature possessed true teeth, have been already alluded to.

In spite of the circumstance that he was unable to discover its real mode of reproduction, the best account that we possess of the habits of the Duck-bill is that one published by Dr. George Bennett, from observations made during the years 1829-1832 in Australia, from which the following extracts are taken:—

After stating that in certain parts of Australia the creature is known to the aborigines by the names of *Mallangong*, or *Tambrit*, the author proceeds to state that its favourite haunts are those tranquil portions of the rivers termed "ponds" by the colonists, on the surface of which various aquatic plants spread their leaves in profusion. Among these plants the Duck-bills may be seen by the cautious observer seeking their food; while the shaded banks afford them excellent situations for the construction of their burrows. "In such circumstances they may be readily recognised by their dark bodies just seen level with the surface, above which the head is slightly raised, and by the circles made in the water around them by their paddling action. On seeing them, the spectator must remain perfectly stationary, as the slightest noise or movement will cause the timid creature instantly to disappear, so acute are they in sight or hearing, or perhaps in both; and they seldom reappear when once frightened. By remaining perfectly quiet, however, when the animal is paddling about, it is possible to obtain an excellent view of its movements in the water; it seldom remains longer than for one or two minutes playing on the surface, but dives, and reappears a short distance above or below the place at which it was observed to descend."

Later on, Dr. Bennett observes that "these creatures are seen in the Australian rivers at all seasons of the year, but are most abundant during the spring and summer months. . . . The best time for seeing them is early in the morning or late in the evening. During floods and freshets they are frequently perceived travelling up and down the rivers; when going down, they appear to allow themselves to be carried by the force of the stream, without making any exertion; but in swimming against the current their muscular power is exerted to the utmost to stem its force, and generally with success. I recollect, however, seeing two make repeated and ineffectual attempts to pass a small waterfall in a rapid part of the river, and, after many persevering efforts, they were unable to attain their object."

In describing the structure of the burrow, the same author states that on a certain occasion on arriving at the spot where one had been partially opened by the natives, he found it situated in a steep bank adjacent to the river, upon which grass and other herbaceous plants grew in abundance. The native guide, by whom he was accompanied, "putting aside the long grass, displayed the entrance of the burrow, distant rather more than a foot from the water's edge. In digging up this retreat, the natives had not laid it entirely open, but had delved holes at certain distances, always introducing a stick for the purpose of ascertaining the direction in which the burrow ran, previously to again digging down upon it. By this method they were able to explore its whole extent with less labour than if it had been entirely laid open. The termination of the burrow was broader than any other part, nearly oval in form; and the bottom was strewn with dry river-weeds, &c., a quantity of which still remained. From this place my sable friend said he had taken last season three young ones, which were about six or eight inches long, and covered with hair. The whole of the burrow was smooth, extending about twenty feet in a serpentine direction up the bank. . . . The burrows have two entrances—one usually at about the distance of a foot from the water's edge, and another under the water. It is, no doubt, by the entrance under the water that the animal seeks refuge within its burrow, when it is seen to dive and not to rise again; and when the poor hunted quadruped is unable to enter or escape from the burrow by the upper aperture, it has recourse to the river-entrance."

As a rule, the burrows of the Duck-bills are situated above the usual level of the river, but do not appear to be out of the reach of the floods which are so common during the winter. At the breeding season the terminal chamber of the burrow is lined with a layer of dried grass and weeds, upon which the white eggs, usually two in number, are deposited. Here the eggs are hatched out by the female parent brooding upon them after the manner of a bird; the pouch on the under surface of her body being never sufficiently large to contain the eggs. After their emergence from the shell the naked and helpless young are nourished by their parent in the manner already mentioned. Although the usual number of eggs is two, it appears that there may be occasionally either three or four. After the period of suckling ceases, the young are said to be fed by the parent on insects and comminuted shell-fish, until sufficiently strong and active to shift for themselves.

Describing the capture of a pair of well-advanced young in the nest, Dr. Bennett states that when first seen, although there was plenty of growling, "there was no movement on the part of the animals to escape. On being taken out, they were found to be full-furred young ones, coiled up asleep, and they growled exceedingly at being exposed to the light of day. There were two of them, a male and female, of the dimensions of ten inches from the extremity of the beak to that of the tail. They had a most beautiful, sleek, and delicate appearance, and seemed never to have left the burrow. The nest, if it may be so termed, consisted of dry river-weeds, the epidermis of reeds, and small dry fibrous roots, strewed over the floor of the cavity, which was of sufficient size to contain the mother and her young."

As we have said, the food of the Duck-bill consists of various small water-animals, such as crustaceans, insects and their larvæ, snails and worms, which are dug out of the soft mud by the tender muzzle, and are taken into the mouth accompanied by a considerable quantity of sand, which results in the rapid wearing away of the molar teeth of the young animal. Sleeping for the greater portion of the day comfortably rolled up in their burrows, the Duck-bills are mainly nocturnal in their habits, not generally issuing forth to feed till the shades of evening proclaim the approach of night, although at times they may be seen abroad in full daylight. Except for the loud growling uttered when they are dug out of their burrows, it appears that they are for the most part silent creatures.

Of the habits of two immature specimens and their dam kept by him for some time in captivity, Dr. Bennett writes as follows:—

"The young animals sleep in various postures, sometimes extended, and often rolled up like a Hedgehog in the shape of a ball. They formed an interesting group, lying in different attitudes in the box wherein I had placed them, and seemed happy and contented. Thus, one was curled up like a dog, keeping its back warm with its flattened tail, which was brought over it; while the other was stretched on its back, its head resting, by way of a pillow, upon the body of the old one, which laid on its side with its back resting against the box,—the delicate beak and smooth clean fur of the young contrasting with the rougher and dirtier appearance of the old one,—all fast asleep. At another time they might be seen—a curious-looking group—one lying on its back with outstretched paws, another on its side, and the third coiled or rolled up like a hedgehog. They shift themselves from one position to another, as they may feel fatigued by lying long in the first; but the favourite posture of the young animals appears to be coiled up like a ball. This is effected by the fore paws being placed under the beak, with the head and mandibles bent down towards the tail, the hind paws crossed over the mandibles, and the tail turned up, thus completing the rotundity. Although furnished with a thick coat of fur, they still seemed particular about being kept warm and comfortable. They would allow me to smooth their fur, but if the mandibles were touched they darted away immediately, these parts appearing to be remarkably sensitive.

"I could permit the young to run about the room as they pleased, but the old specimen was so impatient, and damaged the walls so much by attempts at burrowing, that I was obliged to keep her a close prisoner in the box, where during the day she would remain quiet, huddled up with the young ones, but at night would become very restless, and eager to escape from her place of confinement. A general growl would issue forth from the group if disturbed when asleep."

As indicated in the foregoing extract, the fur of aged individuals becomes frequently very rough and ragged, so that the whole animal presents a dishevelled and almost disreputable appearance, in striking contrast to the sleekness and beauty of the coat of the young before they have left the burrow. Regarding the difference in these respects, between the old and the young Duck-bills, the author above quoted observes that "in the young animals the upper beak was of a similar colour to the same organ in the old specimens; but on its under surface the colour was a beautiful pink, in consequence of the minute blood-vessels being distinctly visible through the delicate epidermis, like the bloom of rosy health on the cheek of an infant. The legs close to the feet were fringed with fine silvery hairs, and the fur on the back, although of a more delicate nature, was similar in colour to that of older specimens; yet the ferruginous hue of the whole extent of the under part of the chest and abdomen had a lighter tinge, dependent probably on the age of the animal."

It may be mentioned in this connection that it has often been considered that there are several species of Duck-bill, but it appears that such supposed species have been founded upon different individual variations of *Ornithorhynchus anatinus*, due to differences of age and peculiarities of coloration. Writing many years ago on this subject, Waterhouse observed that "in the large and mature specimens the fur is crisp and of a dull brown colour; whilst in the smaller specimens it is of a bright brown hue, and soft to the touch, and, on the underparts of the body, is almost white. Those of smaller size and brighter colouring have received the specific names *rufus*, *lævis*, and *brevirostris*; and I may further observe that the specimen from which Dr. Shaw drew up his original account, and which is now in the British Museum collection, presents a similar condition of the fur, and is of small size. The *O. fuscus* of Péron and Lesueur, and the *O. crispus* of Macgillivray, are large specimens, in which the fur is comparatively crisp and of dull colouring." These conclusions are fully confirmed by Mr. Thomas in his Catalogue of the Marsupials and Monotremes in the British Museum.

THE ECHIDNAS. FAMILY ECHIDNIDÆ.

The second family of the Monotremes is represented by the Echidnas, Spiny Ant-eaters, or Porcupine-Ant-eaters, as they are indifferently called, of which there are two or possibly three species, arranged under two genera, one of which is common to Australia, while the other is solely Papuan. The family may be characterised as follows :

Terrestrial and fossorial Monotremes, in which the sexes are not markedly different in size, and the muzzle is in the form of a long, slender, cylindrical, toothless beak, adapted for the protection of the long, worm-like, extensile tongue. Fur thickly mingled with short, stout spines; tail rudimentary; limbs short and subequal, with the toes unwebbed, and furnished with stout and broad claws; soles of feet provided with soft, fleshy cushions, but devoid of pads; palate and tongue spinous. Pouch well-developed during the breeding-season. Hemispheres of the brain marked by numerous convolutions.

The skull is remarkable for its peculiarly smooth and bird-like form; the lower jaw being remarkably long and slender, with the ascending, or coronoid process, as well as the posterior angle, rudimentary.

In habits, the Echidnas differ markedly from the Duck-

PLATE XXXVIII

bills, being exclusively terrestrial, although fossorial, and subsisting solely on ants, which are licked up by the long extensile tongue, after the manner of the Banded Marsupial Anteater. The total loss of the teeth in the Echidnas, in which not even rudiments of these organs have hitherto been detected, suggests that they have been Ant-eaters for a longer period of time than has the Marsupial Ant-eater. As regards their breeding habits, the Echidnas differ from the Duck-bills in that the females carry their eggs—generally two in number—in the pouch, where they are hatched out by the heat of the body.

THE FIVE-CLAWED ECHIDNAS. GENUS ECHIDNA.

Echidna, Cuvier, Tabl. Element., p. 143 (1798).

Form stout and depressed; limbs with five toes, all furnished with claws, those of the fore feet being broad, slightly curved, and directed forwards, while on the hind feet they are more slender, and curved outwards, the second, or second and third, being elongated and considerably exceeding the fourth and fifth in length, while on the first toe, or hallux, the claw is short, blunt, and rounded. Beak about equal in length to the rest of the remainder of the head, and either straight or slightly curved upwards. Tongue tapering at the tip, with its spines restricted to its basal portion.

1. COMMON ECHIDNA. ECHIDNA ACULEATA.

Myrmecophaga aculeata, Shaw, Natur. Miscell., vol. iii., pl. cix. (1792).
Ornithorhynchus hystrix, Home, Phil. Trans., 1802, p. 348.
Echidna hystrix, Geoff., Cat. Mus., p. 224 (1823).
Tachyglossus aculeatus, Illiger, Prodromus Syst. Mamm., p. 114 (1811).
Echidna aculeata, Garnot, N. Bull, Soc. Philom., 1825, p. 45 (1825); Thomas, Cat. Marsup. and Monotr. Brit. Mus., p. 377 (1888).

Echidna australis, Lesson, Hist. Nat. Mamm., vol. v., pl. ii. (1836).
Echidna acanthion, Collett, Forhandl. Vid. Selsk. Christiana, 1884, No. 13 (1885).

(*Plate XXXVIII.*)

VARIETY A.—PORT MORESBY ECHIDNA.

Echidna (*Tachyglossus*) *lawesi*, Ramsay, Proc. Linn. Soc. New S. Wales, vol. ii., p. 32 (1877).
Tachyglossus lawesi, Dubois, Bull. Soc. Zool., vol. vi, p. 268 (1881).

VARIETY B.—HAIRY ECHIDNA.

Echidna setosa, Geoff., Cat. Mus., p. 226 (1803).
Echidna breviaculeata, Tiedemann, Zool., vol. i., p. 592 (1808).
Tachyglossus setosus, Illiger, Prodromus Syst. Mamm., p. 114 (1811).
Echinopus setosus, G. Fischer, Zoogn., vol. iii., p. 694 (1814).

Characters.—General colour of hair black or dark brown; under-parts brown; spines of back long and stout, generally completely concealing the fur, and usually yellow with black tips; tail short and conical, with the extremity naked. Length of head and body varying from about 14 to 20 inches.

Distribution.—From South-Eastern New Guinea, throughout Australia, to Tasmania.

Varieties.—Writing of the different local races of this exceedingly variable and widely-spread Echidna, Mr. Thomas observes that although several so-called species have been founded, and "although the range of variation is very large, yet all the intermediate stages appear to exist between the most widely separated forms. Three geographical races, however, seem to deserve recognition by name—a northern, central, and southern; but these distinguishing characters are too slight and too inconstant to justify their specific distinction."

Of these, the typical or central race, is confined to the mainland of Australia, and is of medium size, attaining a length of about 17 inches, exclusive of the tail. Next to this comes the Port Moresby or Papuan five-clawed Echidna, which is confined to south-eastern New Guinea, and is characterised by its smaller dimensions (length of head and body about 14 inches), and the shorter spines of the back. Curiously enough, the most different from all is the hairy variety, from Tasmania, which attains a length of about 20 inches, and is further characterised by the greater length of the fur, which nearly conceals the spines, and likewise by the elongation of the claw of the third toe of the hind foot, which is nearly equal in length to that of the second.

Habits.—As in the case of the Duck-bill, it has only been comparatively recently proved that the Echidna lays eggs, which are carried about and hatched by the female in the manner alluded to above; we are still, however, in need of much fuller information concerning the breeding-habits of these creatures. In order to enable them to procure with facility their food of ants and their larvæ, Echidnas are provided with very large glands discharging into the mouth, the viscid secretion from which causes the ants to adhere to the long worm-like tongue when thrust into a mass of these insects, after being exposed by the digging powers of the claws of the Echidna's limbs. As a rule, in Australia these animals are found in sandy and rocky districts, and are especially common in the mountains, where they dwell in holes among the rocks. On level ground they move with an unsteady, shuffling gait, with the short and broad fore paws turned inwards, and the claws of the hind feet bent outwards and backwards, so that the inner border of the sole rests on the ground.

Crespuscular and nocturnal in habits, the Echidna, according to Bennett's account, passes most of the day asleep, and displays its great burrowing powers chiefly by night. "Its movements are active," continues this observer, " particularly when engaged in burrowing, which is effected with an extraordinary degree of celerity. When attacked, they roll themselves into a ball similar to the Hedgehog, and, with erect spines, form an excellent defence. They are very restless when in confinement, and pry into every crevice, and if any opening, however small, is found to admit their powerful burrowing fore paws, it will speedily be torn up, and the animal will escape. The only method of carrying the creature is by one of the hind legs, when it may be removed to any place with great facility; for an attempt to seize it by any other part of the body, from its powerful resistance, and the sharpness of the spines, will soon oblige the captor to relinquish his hold, when the animal, rolling itself into a spherical form, is free for some time from aggression. It also resists removal by its power of adhering to any object, as I found on several occasions. When one of these animals was given to me, and placed in the box of the gig to bring home, on arriving there I could not, by any effort, remove it, from its adhering to the boards like a limpet to the rocks (the head and snout being drawn in); only a formidable array of prickles was visible, so sharp, that on the least touch they left a very painful feeling in the hands. So firmly was the animal fixed, that it was impossible to stir it from that position. At last, the method of removing limpets from the rocks was resorted to, and a spade being inserted gradually at one extremity of the animal, it was scraped from its position after some difficulty; and even then it was some length of time before we succeeded in grasping the hind legs and conveying the troublesome creature to the place of confinement allotted to it. So much trouble was given by its burrowing habits and spinal irritation, that its death some time afterwards was not regarded with much regret.

"I have usually observed the animal sleeping rolled up like a ball; when cleaning itself, it uses only the hind claws, placing itself in various positions, so as to be enabled to reach the part of the body to be operated on. I never heard a sound of any kind uttered by this animal."

Captive Echidnas have been fed at first upon the pupæ of ants and milk, but have subsequently been accustomed to eat chopped egg and meat. In drinking, which they do frequently and largely, the fluid is licked or lapped up by a rapid protrusion and retraction of the worm-like tongue.

Like all burrowing animals, the Echidna has the humerus, or upper bone of the arm of great width, and furnished with a number of strong ridges and crests necessary for the attachment of the powerful muscles required to move the claws in the work of excavating the ground. Although there is a certain family likeness in that bone among all such burrowing Mammals, yet the humerus of the Echidna is markedly different from that of all other members of the class (save in a certain degree, that of the Duck-bill); and it is remarkable that this curious type of humerus is most nearly approached by the aforesaid extinct Anomodont reptiles, which there is good reason to believe are nearly allied to the ancestors of the Monotremes.

THE THREE-CLAWED ECHIDNAS. GENUS PROECHIDNA.

Proechidna, Gervais, Octéographie Monotrem., p. 43, 1877.

Usually only three claws on both fore and hind feet, but the first and fifth toes represented by several joints, and in some cases with fully developed claws. In the hind foot the claws decrease regularly in length from the second to the fourth toe. Beak much elongated and curving down, its length being nearly equal to twice that of the rest of the head.

Owing to the circumstance that an apparently abnormal individual of the typical species of this genus has been found with five anterior and four posterior fully developed claws, it will be apparent that the main ground for referring these Spiny Ant-eaters to a genus apart from *Echidna* is the much greater proportionate length of the beak, although there is also a difference in the number of the vertebræ. At best, however, these differences are but trivial, and, were it not for the inconvenience of perpetually changing accepted systems of classification, the writer considers that it would be preferable to include the whole of the members of the family in a single genus.

The three-clawed Echidnas are restricted to New Guinea.

I. BRUIJN'S ECHIDNA. PROECHIDNA BRUIJNI.

Tachyglossus bruijnii, Peters and Doria, Ann. Mus., Genova, vol. ix., p. 183 (1876).
Acanthoglossus bruijnii, Gervais, Comptes Rendus, 1877, p. 837.
Proechidna bruijnii, Gervais, Ostéographie Monotrem. p. 43 (1877); Thomas, Cat. Marsup., Brit. Mus., p. 383 (1888).
Proechidna villosissima, Dubois, Bull. Mus., Belg., vol. iii., p. 110 (1884).
Echidna bruijnii, Flower, Cat. Osteol. Mus., R. Coll. Surgeons, pt. ii., p. 753 (1884).

Characters.—About equal in size to the largest specimens of the Tasmanian race of *Echidna aculeata*. Fur thick, coarse, and woolly, sparsely intermingled with flattened bristles; in colour, the fur uniform dark brown or black, sometimes becoming nearly white on the head. Spines on back much shorter, as well as fewer than in *Echidna;* their colour being generally entirely white, although their bases may, in some cases, be brown. Length of head and body from about 19 to 20 inches.

Distribution.—North-western New Guinea (Arfak Mountains).

Although, as in the case with Papuan Mammals in general, little or nothing is known of the habits of this species, these are in all probability very similar to those of the common Echidna; it being reported by the natives that the present species, like the latter, dwells in crevices and holes among rocks.

II. BLACK-SPINED ECHIDNA. PROECHIDNA NIGRO-ACULEATA.

Proechidna nigro-aculeata, Rothschild, Proc. Zool. Soc., 1892, p. 545.

Characters.—The specimen on which this reputed species was founded is described as differing from the last by its much larger size, extremely robust limbs, and much shorter claws. Moreover, instead of the dense woolly brown hair of *P. bruijni*, among which are embedded the few and scattered spines, in the present form the hair is long, bristly, and very sparingly sprinkled over the body, the legs being almost naked. Then, again, while in *P. bruijni* the spines are few, short, thin, and white, in *P. nigro-aculeata* they are nearly as numerous as in the Common Echidna, and are likewise of great length and thickness, as well as being of a jet black colour. The claws of this form are further distinguished by their relative shortness and breadth, as well as by the excavation of their inferior surfaces; while the tail is longer and stouter than in the typical representative of the genus. The total length of the type specimen of this species is upwards of 31 inches, against 24 inches in the largest individual of *P. bruijni*.

Habits.—Western New Guinea (Charles Lewis Mountains).

Additional specimens are required before it can be definitely asserted that this form indicates anything more than a large race of *P. bruijni*, characterised by the abundance and dark colour of its spines; since its points of difference from the ordinary form do not appear to be greater than those distinguishing the local races of the Common Echidna.

PART III.

EXTINCT MARSUPIALS AND MONOTREMES.

INTRODUCTORY.

In addition to remains belonging to existing species, the superficial deposits and caverns of Australia have yielded evidence of the existence of a number of Marsupials belonging to species, genera, or even families entirely distinct from all their living allies, and in many cases attaining dimensions far exceeding any of those of the latter. All these extinct Marsupials belong, however, to the same general groups as those now inhabiting Australia; and they for the most part appear to have existed during the Pleistocene, or latest geological period, that is to say, to the time when the Mammoth and Woolly Rhinoceros flourished in Europe.

In South America we have likewise evidence that during the same period species of Opossum were abundant, at least a large number of which were identical with those still inhabiting the same country. When, however, we examine the older Tertiary strata of Patagonia we find remains of extinct Marsupials which appear, so far as the present evidence may be credited, to be more nearly allied to the Australian *Dasyuridæ*, than to the Opossums. Still earlier in the Tertiary period, namely in the upper portion of its Eocene, or Oligocene, division there is abundant evidence that Opossums, apparently inseparable from the existing genus *Didelphys*, were spread not only over North America, but likewise over a large portion of Europe.

Passing downwards to the strata lying below the Tertiaries, that is to say those equivalent to or older than the chalk, and collectively forming the great Secondary system, Marsupials have been found both in Europe and North America; all of which appear to belong to, or to be more or less nearly allied to the Polyprotodont sub-division of the order; the Diprotodonts, as we have already had occasion to mention, being quite unknown beyond the limits of the Australasian region. Such primitive Secondary Marsupials occur not only in the Cretaceous strata, or those equivalent to our chalk and greensands, but likewise range through the subjacent series of Jurassic or Oolitic rocks, and are sparingly represented in the still older beds of the Trias.

As regards the Monotremes, we have no evidence of their existence previous to the Pleistocene beds of Australia, where remains referable to two of the existing genera have been discovered. As we have already mentioned, it is, however, possible that those remarkable older Tertiary and Secondary Mammals known as the Multituberculata may indicate a group allied to the Monotremes, but if so, probably forming a separate and distinct sub-ordinal group of the sub-class Prototheria.

In the following brief notice of the fossil Marsupials and Monotremes, all mention will be omitted of such as are identical with existing species; while, from the limits of space, and the imperfection of our acquaintance with the affinities of many members of the former group, only the more interesting and better known types will be alluded to at all.

FAMILY MACROPODIDÆ (*suprà* p. 11).

GENUS PALORCHESTES.

Palorchestes, Owen, Phil. Trans., 1874, p. 797.

PALORCHESTES AZAEL.

Palorchestes azael, Owen, *op. cit.*, p. 798; Lydekker, Cat. Foss. Mamm. Brit. Mus., pt. v., p. 237 (1887).

Characters.—This species, the sole representative of its genus, is by far the largest known member of the Kangaroo family; the length of the complete skull being somewhere about sixteen inches. As a genus, this gigantic Kangaroo is characterised by the two branches of the lower jaw being firmly welded together at their union; so that the scissor-like action occurring between the lower incisor teeth of existing Kangaroos was impossible. The union between the two branches of the lower jaw is elongated; while the interval between the incisor and premolar teeth was likewise of considerable length. The last upper premolar tooth, which is of medium length, has a well-developed lobe on its inner side, and thereby assumes a triangular form, thus totally departing from the cutting type of tooth characterising the existing members of the family. The corresponding lower tooth is of the same general type, but has its additional lobe on the outer, instead of the inner side. The molar teeth have smooth enamel, but lack any basal ledge in front of the first ridge, thus resembling the corresponding teeth of the existing *Macropus magnus;* in the lower molars the median longitudinal bridge connecting the two transverse ridges is well developed. In spite of the union of the two branches of the lower jaw, the lower pair of incisor teeth had the same elongated and spatulate form which characterises existing Kangaroos.

Notwithstanding its gigantic stature, it may be inferred from the similarity of the limb-bones of all the fossil Kangaroos to those of the living species that this extinct form progressed by leaps in the same manner as its modern cousins.

Distribution.—Victoria, Queensland, and New South Wales; the remains occurring both in the superficial deposits of the two former districts, and likewise in the caverns of the Wellington Valley in the latter.

GENUS PROCOPTODON.

Procoptodon, Owen, Phil. Trans., 1874, p. 788.

The Kangaroos of this genus, although smaller than the preceding, were likewise, in most cases, of very large size. Agreeing with the latter in the structure of their premolar teeth, and likewise in the bony union between the two branches of the lower jaw of the adult, they differed by the extreme shortness and depth of the whole lower jaw, and the smallness of the interval between the incisor and premolar teeth. Another peculiarity is to be found in the circumstance that the enamel of the molar teeth is generally thrown into a number of vertical folds and puckerings; while the lower incisor teeth differ from those of all living Kangaroos in their cylindrical form. The skull differs from that of *Palorchestes* in having unossified vacuities in the palate, as in the existing members of the family. The lower jaw is easily recognised by the conversion of the usual pit on the outer side of its hinder portion into a complete pocket, owing to the great development of its outer wall.

Certain limb-bones which have been tentatively assigned to this genus, indicate, if their association with the jaws be correct, that the disproportion between the front and hind limbs was less marked than is the case in *Macropus*.

I. PROCOPTODON GOLIAH.

Macropus goliah, Owen, in Waterhouse's Nat. Hist. Mamm., vol. i., p. 59 (1846).
Procoptodon goliah, Owen, Phil. Trans., 1874, p. 783; Lydekker, Cat. Foss. Mamm. Brit. Mus., pt. v., p. 234 (1887); *id.* Quart. Journ. Geol. Soc., vol. xlvii., p. 571 (1891).

Characters.—A large species characterised by the slenderness and length of the lower jaw, and by the vertical folds on the enamel of the molars being very strongly developed.

Distribution.—Queensland and New South Wales.

II. PROCOPTODON RAPHA.

Procoptodon rapha, Owen, Phil. Trans., 1874, p. 78; Lydekker, Cat. Foss. Mamm. Brit. Mus., pt. v., p. 235 (1887); *id.* Quart. Journ. Geol. Soc., vol. xlvii., p. 571 (1891).

Distinguished from the preceding by the shorter and deeper lower jaw.

Distribution.—Queensland and New South Wales.

III. PROCOPTODON OTUEL.

Pachysiagon otuel, Owen, Phil. Trans., 1874, p. 784.
Procoptodon otuel, Lydekker, Cat. Foss. Mamm. Brit. Mus., pt. v., p. 236 (1887).

A species generally of smaller size than either of the others, from which it is distinguished by the almost complete absence of vertical foldings in the enamel of the molar teeth.

Distribution.—Queensland.

GENUS STHENURUS.

Sthenurus, Owen, Phil. Trans., 1874, p. 264.

STHENURUS ATLAS.

Macropus atlas, Owen, in Mitchell's Australia, vol. ii., p. 359 (1838).

Protemnodon anak (*in parte*), Owen, Phil. Trans., 1874, p. 275.

Sthenurus atlas, Owen, Phil. Trans., 1874, p. 264; Lydekker, Cat. Foss. Mamm. Brit. Mus., pt. v., p. 232 (1887).

Characters.—This genus and species, while agreeing with *Procoptodon* in the characters of the last premolar tooth, differs in having by the two branches of the lower jaw united only by ligament, and in the spatulate form of the lower incisor teeth, which resemble those of *Macropus*, and probably had a similar scissor-like action against one another. The molar teeth are devoid of vertical foldings of the enamel, and are very short and wide; the longitudinal bridge connecting their two transverse ridges being very imperfect, and the anterior basal ledge of the upper molars being unconnected by such a bridge with the first transverse ridge. In size the species was large, the skull being probably about a foot in length.

This genus forms a connecting link between the existing *Macropus* and the extinct *Procoptodon*.

Distribution.—Queensland and New South Wales.

GENUS MACROPUS (*suprà*, p. 14).

1. MACROPUS TITAN.

Macropus titan, Owen, in Mitchell's Australia, vol. ii., p. 360 (1838); Lydekker, Cat. Foss. Mamm, Brit. Mus., pt. v., p 225 (1887).

Characters.—Resembling the existing *Macropus giganteus* (to which it is in all respects closely allied) in the extremely small size of the last premolar tooth. This species is distinguished by its superior dimensions, and by the very general occurrence of one or more vertical grooves on the hinder aspect of the lower molars; but it is not improbable that the fossil and living forms pass imperceptibly one into the other.

Distribution.—Australia.

II. MACROPUS FERRAGUS.

Pachysiagon ferragus, Owen, Extinct Mamm. Australia, p. 449 (1879).
Macropus ferragus, Lydekker, Cat. Foss. Mamm. Brit. Mus., pt. v., p. 230 (1887).

An imperfectly known species, allied to the preceding, but of somewhat larger size.

Distribution.—Queensland and New South Wales.

III. MACROPUS ALTUS.

Phascolagus altus, Owen, Phil. Trans., 1874, p. 261.
Macropus altus, Lydekker, Cat. Foss. Mamm. Brit. Mus, pt. v., p. 223 (1887).

Characters.—This large species is nearly allied to the existing *Macropus robustus*, in which the upper molars have no longitudinal bridge between the anterior basal ledge and the first transverse ridge; the last premolar being small, and (as in *M. giganteus*) frequently shed in the adult. To *M. robustus* the present species appears to bear a relation similar to that held by *M. titan* to *M. giganteus*.

Distribution.—New South Wales and Queensland.

IV. MACROPUS COOPERI.

Osphranter cooperi, Owen, Phil. Trans., 1874, p. 261.
Macropus cooperi, Lydekker, Cat. Foss. Mamm. Brit. Mus., pt. v., p. 224 (1887).

An imperfectly known species nearly related to *M. robustus*, but distinguished from it, as well as from *M. altus*, by the much slighter downward inclination of the upper border of the lower jaw in front of the premolar teeth.

Distribution.—Queensland.

V. MACROPUS BREHUS.

Sthenurus brehus, Owen, Phil. Trans., 1874, p. 272.
Protemnodon mimas, Owen, *op. cit.*, p. 278.
Macropus brehus, Lydekker, Cat. Foss. Mamm., pt. v., p. 207 (1887).

Characters.—This and the three remaining extinct species of the genus belong to the group of larger Wallabies, as defined on p. 14, although they vastly exceed all the existing species of that group in point of size, being, indeed, superior in this respect to the largest living Kangaroos. In all of them the last premolar is characterised by its great longitudinal length, as well as by the vertical flutings on the sides of its sharp and trenchant crown. The upper molars have no longitudinal bridge connecting the anterior basal ledge with the first transverse ridge; and in both jaws the molars are relatively wide, and have a low longitudinal bridge connecting the two transverse ridges.

By Garrod these gigantic extinct Wallabies were regarded as allied to the Kangaroos of the genus *Dorcopsis*, which they resemble in the length and general form of the last premolar. The elongation of the latter tooth is, however, not so great as in that genus, from which these Wallabies are further distinguished by the presence of the longitudinal bridge connecting the transverse ridges of their molars, by the large size of the innermost pair of upper incisor teeth, and by the total absence of the upper canine.

In this the largest species the estimated length of the skull is about twelve inches, while it is further characterised by the length of the last premolar considerably exceeding that of the first tooth of the molar series. The lower molars show a distinct posterior basal ledge, and the innermost pair of upper incisors are much larger than either of the others.

Distribution.—New South Wales and Queensland.

VI. MACROPUS RÆCHUS.

Protemnodon ræchus, Owen, Phil. Trans., 1874, p. 281.
Protemnodon antæus, Owen, Extinct Mamm. Australia, p. 448 (1877).
Macropus ræchus, Lydekker, Cat. Foss. Mamm. Brit. Mus., pt. v., p. 212 (1887).

Closely allied to the last, of which, indeed, it may be only a variety distinguished by certain peculiarities in the lower teeth.

Distribution.—Queensland.

VII. MACROPUS ANAK.

Macropus anak, Owen, Proc. Geol. Soc., vol. xv., p. 185 (1859); Lydekker, Cat. Foss. Mamm. Brit. Mus., pt. v., p. 214 (1887).
Protemnodon anak, Owen (*in parte*), Phil. Trans., 1874, p. 275.
Protemnodon og, Owen (*op. cit.*), p. 227
Sthenurus atlas, Owen (*in parte*), *op. cit.*, p. 265.

Characters.—Closely allied to *M. brehus*, but generally of smaller size, and frequently without the posterior basal ridge to the last molar of the lower jaw.

Distribution.—Queensland, South Australia, and New South Wales.

VIII. MACROPUS MINOR.

Sthenurus minor, Owen, Proc. Zool. Soc., 1877, p. 353.
Macropus minor, Lydekker, Cat. Foss. Mamm. Brit. Mus., pt. v., p. 218 (1887).

Distinguished from the last species by its considerably smaller dimensions, the shorter last premolar, and the proportionately narrower upper molars.

Distribution.—New South Wales and Queensland.

GENUS TRICLIS.

Triclis, De Vis, Proc. R. Soc., Queensland, ser. 2, vol. iii., p. 8 (1888).

I. TRICLIS OSCILLANS.

Triclis oscillans, De Vis, *loc. cit.*

Founded on a lower jaw from the superficial deposits of Queensland, which apparently belonged to an animal allied to *Hypsiprymnodon*, but of much larger dimensions. The fossil agrees with the living form in having the last premolar vertically grooved and inclined outwards, but differs, as it also does from every other member of the present family, in the presence of an additional minute tooth between the last premolar and the incisor, this additional tooth probably representing the canine, but being possibly a premolar. In the presence of this tooth *Triclis* closely connects the *Macropodidæ* with the *Phalangeridæ*

FAMILY PHALANGERIDÆ (*suprà* p. 75).

In addition to remains of several of the existing representatives of the family, certain bones are regarded as indicating the existence of extinct genera and species more or less nearly allied to the living ones; all such bones having been obtained from the superficial deposits of Queensland. Among these, one fragment is regarded as belonging to a large extinct species of the existing genus *Pseudochirus*; while a second is referred to an allied but extinct genus termed *Archizonurus*; and a third is assigned to *Phalanger*. More interesting still are certain specimens believed to belong to a large animal allied to the living Koala, but, from certain supposed generic differences, named *Coalæmus*. Far more satisfactory than the above-mentioned fragmentary specimens, are the well preserved remains of the following genus, which indicates a distinct sub-family (*Thylacoleontinæ*) of the *Phalangeridæ*:

GENUS THYLACOLEO.

Thylacoleo, Owen, in Gervais's Zool. et Pal. Françaises, 1st ed., pt. i., p. 192 (1849-52).

1. THYLACOLEO CARNIFEX.

Thylacoleo carnifex, Owen, Phil. Trans., 1859, p. 309; Lydekker, Cat. Foss. Mamm., Brit. Mus., pt. v., p. 189 (1887).

The single representative of this genus was a gigantic animal allied to the Phalangers, but distinguished by the peculiar and specialised character of its dentition; the skull measuring some eight or ten inches in length, and being remarkable for its breadth, rounded form, and general massiveness, and for the socket of the eye being completely surrounded by bone. There are altogether thirty-two teeth, the first pair of incisors in the upper jaw being of great size, and the canine and two anterior premolars small and functionless. In the lower jaw there is a similar large single pair of first incisor teeth, which are directed more upwardly than in existing members of the family; behind these come two minute premolars, followed by an enormous cutting last premolar, which is of the same type as the corresponding tooth of the existing Rat-Kangaroos, although of proportionately much larger dimensions, and biting against a similar tooth in the upper jaw. The molar teeth (one pair in the upper, and two in the lower jaw) which succeed these large premolars posteriorly are small, rounded, and of no functional importance.

The remarkably trenchant form of the last premolar tooth in this strange extinct representative of the Phalangers not unnaturally led to the conclusion that the creature was a carnivore preying upon the large herbivorous Marsupials which were its contemporaries; and it accordingly received the specific name which it still bears. Fuller acquaintance with its anatomy revealed, however, its intimate kinship with the Phalangers; and when this was fully realised, it was argued that *Thylacoleo* must be purely a vegetable eater. Many of the Cuscuses are, however, partly carvivorous in their habits; and in our own opinion it seems probable that in this respect their gigantic extinct cousin resembled them to a certain extent. Not that we mean to assert that *Thylacoleo* was a creature which preyed on large Mammals, since to attack and overcome such its teeth are clearly not suited; but we do think that it may have probably killed and devoured the smaller Mammals, as well as such birds as it was able to catch. To our thinking, the huge last premolar tooth does not appear one well adapted for an exclusively vegetable diet, unless indeed it were employed to slice up large bulbous roots after the manner of a turnip-cutting machine. Even, however, if roots of this nature existed in Australia during the time that *Thylacoleo* flourished, we fail to see how the creature could have introduced them whole into its mouth and sliced them up in this manner. That there must have been some special use for such extraordinary teeth is pretty clear, as otherwise they would never have been likely to have been evolved; but it still remains for the investigator of the history of the animals of a past epoch to declare what this use was. Of the bones of the limbs we are, unfortunately still ignorant.

Distribution.—Australia generally.

FAMILY DIPROTODONTIDÆ.

GENUS DIPROTODON.

Diprotodon, Owen, in Mitchell's Australia, vol. ii., p. 362 (1838).

1. DIPROTODON AUSTRALIS.

Diprotodon australis, Owen, *loc. cit.*; Lydekker, Cat. Foss. Mamm. Brit. Mus., pt. v., p. 170 (1887).

This, the largest of known Marsupials, is the representative not only of a distinct genus, but likewise of a separate family. Having an elongated skull of about a yard in length, this enormous Marsupial may be compared in size to the largest Rhinoceros, although it probably stood taller on the legs.

The dentition corresponds generally to that of the Kangaroos, that is to say there are no canine teeth, and the incisors, of which there are three pairs in the upper, and one in the lower jaw, are separated by a long gap from the grinding teeth; the latter comprising one pair of premolars and three pairs of molars in each jaw. The upper incisor teeth decrease in size from the first to the third, each pair being in contact in the middle line of the palate; and the first pair being very large and chisel-like, with a coating of enamel on their front surfaces only, and growing continually throughout the life of their owner. The single pair of lower incisors are very large, directed almost straight forwards, and nearly cylindrical in section, thus differing very widely from the corresponding spatulate teeth of the Kangaroos. Each of the grinding teeth carries a pair of bold transverse ridges, very similar to those of the molars of the Kangaroos, but not connected by a median longitudinal ridge. The lower jaw differs from that of the *Macropodidæ* in the absence of any pit on the outer side of the posterior portion; and the palate also differs from that of the existing members of the same family in being completely ossified.

That such an enormous and bulky animal could not have hopped after the manner of a Kangaroo is self-apparent; and this is borne out by the characters of the limb-bones, which show that the fore and hind legs were of normal relative proportions. Until quite recently the structure of the feet was not fully known, but from a letter from Dr. E. C. Stirling published in the "Proceedings of the Zoological Society," it appears that the hind foot was furnished with five toes, and it is probable that the front pair were similarly provided. In the living state these toes were probably encased either in hoofs, or furnished with large broad nails. The tail appears to have been relatively short, measuring only a little over a foot in length, and was therefore utterly unlike that appendage in the Kangaroo.

Till a short time ago, we were acquainted with the *Diprotodon* only by isolated skulls, jaws, teeth, and bones; but in the letter just referred to Dr. Stirling announces the discovery of hundreds of complete skeletons in a salt-lagoon situated some twenty or thirty miles north of a still larger dry salt-lagoon in South Australia known as Lake Frome, and situated about six hundred miles to the northward of Adelaide. The deposits in which this hecatomb of remains are embedded is referred to the Pliocene period by the Australian geologists, or the one immediately preceding the Pleistocene.

According to Dr. Stirling's letter, Lake Mulligan, in which the remains occur, is a relatively small lagoon of about eight miles in diameter; the skeletons being situated about midway between the east and west shores, and lying two or three feet below the surface. "Usually," says the letter, "the salt-crust is not firm enough for bullock traffic, and I may safely say that thousands of bullocks have at different times been bogged in crossing or attempting to cross. .... It would appear that an immense herd of these [Diprotodons] and other animals had got bogged, probably in seeking water in a dry season, just as cattle do now in the north by hundreds. There is every indication that all this region of South Central Australia was formerly occupied by fresh-water lakes. We have, for instance, remains of Alligators or Crocodiles from a district not far off, and other indications of fresh-water life. Of course we are on the look-out for *Thylacoleo*; but, so far, it does not appear to have been met with; but I am quite hopeful that if we can manage to prosecute the search," such remains will ultimately be discovered.

The difficulties of transport for such a great distance across an arid and inhospitable country are, of course, enormous, but when these are overcome, as we trust they will be, and some at least of the skeletons safely housed in the Museum at Adelaide, zoologists will look anxiously for their description, by which our knowledge of the bony structure of this strange monster will be rendered complete.

Distribution.—Australia generally.

In general structure the *Diprotodontidæ* appear to connect the *Phalangeridæ* with the undermentioned extinct family.

FAMILY NOTOTHERIIDÆ.

GENUS NOTOTHERIUM.

Notothcrium, Owen, Cat. Foss. Mamm. Aves. Mus. R. Coll. Surgeons, p. 314 (1845).

1. NOTOTHERIUM MITCHELLI.

Notothcrium mitchelli, Owen, *loc. cit.*; Lydekker, Cat. Foss. Mamm. Brit. Mus., pt. v., p. 162 (1887).

Nototherium inerme, Owen, *op. cit.*, p. 316.
Nototherium victoriæ, Owen, Phil. Trans., 1872, p. 61.

The genus *Nototherium*, in which it is difficult to recognise more than a single species, appears to form a connecting link between the preceding family and the Wombats, the skull, limb-bones, and vertebræ being nearest to those of the latter, the lower jaw presenting characters intermediate between the two, while the molar teeth are similar to those of the former.

Of somewhat smaller size than the *Diprotodon*, the *Nototherium* may be distinguished at a glance by its extremely short and wide skull, in which the region of the nose is curiously upturned. In number, the teeth agree with those of the *Diprotodon*, but the upper incisors are of moderate size, not chisel-shaped, and the second and third pairs are separated in the middle line. Although the upper molars have no longitudinal bridge connecting the two transverse ridges, an incomplete one is developed in those of the lower jaw, and when un-worn the ridges of these teeth sometimes tend to a crescent form. The lower jaw has its inferior border highly convex, and its two branches welded together by bone at their junction in front.

The humerus, or upper arm-bone, tentatively assigned to this animal is very like that of the Wombats; but there is a possibility that it may belong to a gigantic extinct representative of the latter. Although very imperfectly known, the feet also seem to approximate in structure to those of the latter group.

If the humerus assigned to it be rightly referred, it would appear that, in spite of its gigantic dimensions, the *Nototherium* was an animal of burrowing, or at all events digging, habits. The structure of its molar teeth is such as might easily be modified into the more specialised type characteristic of the Wombats, so that it is conceivable that both the latter and the animal under consideration may have been evolved from a common ancestor.

FAMILY PHASCOLOMYIDÆ (*suprà*, p. 122).

GENUS PHASCOLOMYS (*suprà*, p. 124).

In addition to remains of the two existing forms, several extinct species of Wombat have been described from the superficial deposits of Australia, but since these do not present any features of special interest, they will not be further mentioned here.

GENUS PHASCOLONUS.

Phascolonus, Owen, Phil. Trans., 1872, p. 257.

PHASCOLONUS GIGAS.

Phascolomys gigas, Owen, Encyclopædia Britannica, 8th ed., vol. xvii., p. 175 (1859).
Phascolonus gigas, Owen, Phil. Trans., 1872, p. 257; Lydekker, Cat. Foss. Mamm. Brit. Mus., pt. v., p. 157 (1887).

This species was far larger than any existing Wombat, and may be compared in size to a Tapir. It is at present doubtful if it be really entitled to generic separation from *Phascolomys*, the writer having once urged the necessity for such separation in the belief that the teeth mentioned under the next heading belonged to it.

Dr. Stirling states that several skeletons, probably belonging to this species, which he compares in size to a Bullock, have been disinterred in the salt-deposit in South Australia, mentioned under the head of *Diprotodon*. It is to be hoped that these specimens will definitely decide the question as to the nature of the upper incisors; the lower ones are very similar to those of existing Wombats, and, according to De Vis, the same is true with regard to the upper ones.

Distribution.—Queensland, New South Wales, and South Australia.

GENUS SCEPARNODON.

Sceparnodon, Owen, Phil. Trans., 1884, p. 245.

1. SCEPARNODON RAMSAYI.[1]

Sceparnodon ramsayi, Owen, *loc. cit.*; Lydekker, Proc. Royal Soc., vol. xlix., p. 60 (1891); De Vis, Proc. Linn. Soc. New S. Wales, ser. 2, vol. viii., p. 11 (1893).

This genus and species were established on the evidence of large chisel-like incisor teeth, remarkable for their breadth, flatness, and thinness, some examples being fully an inch and a half in width, and not more than a quarter of an inch in thickness. In a clay deposit at Bingera, on the frontier of New South Wales, numbers of these teeth are found separately, and likewise numbers of the jaws of *Phascolonus gigas* without the incisors. Seeing that there were no molar teeth known which could be assigned to *Sceparnodon*, while the upper incisors of *Phascolonus* were unknown, the author contended that the remains described under these two names pertained to one and the same animal. Recently, however, Mr. De Vis has stated that he has obtained the upper incisor of *Phascolonus*, which is quite different from the teeth of the *Sceparnodon*. If his determination be certain, the two genera must be distinct, although there still remains the difficulty that no molar teeth are known which can be assigned to the latter. If, however, his identification is to be accepted, it appears, as already mentioned, very doubtful if *Phascolonus gigas* can be generically separated from *Phascolomys*.

Distribution.—Queensland, New South Wales, and South Australia.

FAMILY DASYURIDÆ (*suprà*, p. 150).

GENUS THYLACINUS (*suprà*, p. 150).

THYLACINUS SPELÆUS.

Thylacinus spelæus, Owen, Cat. Foss. Mamm. Aves. Mus. Roy. Coll. Surgeons, p. 335 (1845); Lydekker, Cat. Foss. Mamm. Brit. Mus., pt. v., p. 264 (1887).

Thylacinus major, Owen, Extinct Mamm. Australia, p. 106 (1877).

Distinguished from the existing Tasmanian species by its considerably larger dimensions. Small specimens of the skull and jaws cannot be distinguished from those of large males of the latter; but it is probable that such specimens indicate female individuals, since the sexual disparity of size was probably as well marked as in the living form

Distribution.—New South Wales and Queensland.

GENUS SARCOPHILUS (*suprà*, p. 153).

SARCOPHILUS LANIARIUS.

Dasyurus laniarius, Owen, in Mitchell's Australia, vol. ii., p. 363 (1838).

Sarcophilus laniarius, Lydekker, Cat. Foss. Mamm. Brit. Mus., pt. v , p. 265 (1887).

Distinguished from the existing *S. ursinus* in much the same manner as *Thylacinus spelæus* differs from *T. cynocephalus*, although the disparity between the size of the extinct and living forms is rather less marked than in the case of the latter. In the fossil species the pits between the upper molar teeth are slightly deeper than in the living one; and there is stated to be some difference in the form of the unossified spaces in the palate, and in the contour of the lower jaw. As in the case of the *Thylacines*, the existing Tasmanian Devil may probably be regarded as the direct descendant of the extinct continental form, dwarfed by the smallness of the area it inhabits.

Distribution.—New South Wales and Queensland.

GENUS PROTHYLACINUS.

Prothylacinus, Ameghino, Rev. Argent. Hist. Nat., vol. i., p. 312 (1892).

This and the following genera from the (Miocene) Tertiary of Patagonia are provisionally included in the present family, where they are placed by their founder. They appear to be more or less closely allied to *Thylacinus*, and to have no near relationship to the American *Didelphyidæ*.

Having the same number of teeth as *Thylacinus*, and exhibiting a similar inflection of the angle of the lower jaw, the present genus appears to be mainly distinguished by the lower premolar teeth being less closely approximated to one another.

PROTHYLACINUS PATAGONICUS.

Prothylacinus patagonicus, Ameghino, Rev. Argent. Hist. Nat., vol i., p. 312 (1892).

The single species of the genus appear to have been approximately equal in point of size to the existing Thylacine.

Distribution.—Tertiary deposits of Patagonia.

GENUS AMPHIPROVIVERRA.

This name has been proposed by Ameghino to replace *Protoproviverra*, which had been previously employed for a totally different genus of extinct Mammals, but the author is at present unaware where it was first published.

The characters by which this genus is distinguished from the preceding are not very clearly defined. It appears, however, to be characterised by the small size of the incisor teeth (of which there are four pairs in the upper, and three in the lower jaw, as in the existing *Dasyuridæ*); and by the laterally compressed canines and premolars, the latter having a small posterior basal tubercle. The palate has large unossified vacuities, and the inflection of the angle of the lower jaw is well-marked.

AMPHIPROVIVERRA MANZANIANA.

Protoproviverra manzaniana, Ameghino, Rev. Argent. Hist. Nat., vol. i., p. 312 (1892).

This species is of considerably smaller size than the representative of the last genus, and may apparently be roughly compared in size with *Sarcophilus ursinus*.

Distribution.—Tertiary formation of Patagonia. Remains from the same deposits have been regarded as indicating other species of the present genus.

GENUS PERATHEUTES.

Peratheutes, Ameghino, Rev. Argent. Hist. Nat., vol. i., p. 313 (1892).

An imperfectly known genus, comprising relatively small forms founded on the evidence of specimens of the lower jaw, in which the number of the incisor teeth is not shown. These are said to be characterised by the uninterrupted series formed by the premolar and molar teeth, the small size of the latter, and the extreme compression of the canine.

1. PERATHEUTES PUNGENS.

Peratheutes pungens, Ameghino, Rev. Argent. Hist. Nat., vol. i., p. 313 (1892).

the type species of the genus, several others having been named.

Distribution.—Tertiary formation of Patagonia.

Before quitting these interesting, but still very imperfectly known, extinct Patagonian Marsupials, a few words may be added as to their importance from a distributional point of view assuming that (as appears to be justified by the number of their incisor teeth) they are rightly regarded as allied to the *Dasyuridæ*. In the first place, it may be mentioned that it has long been a matter of common knowledge that there exist certain very remarkable relationships between the fauna and flora of all the great southern continents. For instance, among Mammals, the rodent family *Octodontidæ* is peculiar to South (including Central) America and Africa, including Algeria. Then, again, among fishes, the family of the *Chromidæ* is confined to the rivers of South America and Africa, with one outlying genus in India; while the True Mud-fishes (*Lepidosiren* and *Protopterus*) are solely South American and Ethiopian, the third representative of the same family being the Baramunda (*Ceratodus*) of Queensland. Again, the connection between the flora of Africa and that of Western Australia is so intimate as to have induced Mr. Wallace to express his belief that there must have been some kind of land connection, although not necessarily a continuous one, between these two widely distant areas. The connection between the fauna of India and that of Ethiopian Africa is now too well known to stand in need of comment. The matter does not, however, end here; for if we go back to the Secondary epoch there are equally striking evidences of the connection between the animals and plants of the southern continents. For instance, an extinct Saurian genus known as *Mesosternum*, which appears to have been allied to the Plesiosaurs of the Lias, is known from early Secondary strata in

Brazil and South Africa, and nowhere else. Then, again, the remarkable Anomodont reptiles (*Dicynodon*, &c.) of South Africa are closely connected with those of India; while the respective alliances between the extinct Labyrinthodont Amphibians and the Secondary floras of South Africa, India, and Australia are too well known to need more than mention.

It appears, then, that, altogether apart from the extinct Marsupials, the common factors connecting the faunas and floras of the four great southern prolongations of the continental land of the globe undoubtedly point, not only to a more or less intimate connection between those several areas, but also to their more or less partial isolation from the more northern lands.

FAMILY DIDELPHYIDÆ (*suprà*, p. 193).
GENUS DIDELPHYS (*suprà*, p. 196).

The number of species of Opossums which have been recorded in a fossil state is so large that it would be wearisome to detail them here, more especially as the teeth and jaws of all the members of the family are very similar in structure, and consequently difficult to distinguish from one another in a fossil state. It will consequently suffice to notice some of the chief formations in which remains of the genus occur.

In the first place, remains of a large number of species of Opossums are found in the Pleistocene deposits of the caverns of Minas Geraes, Lagoa Santa, inland from Rio de Janeiro, Brazil. The greater number of these remains appear inseparable from species still inhabiting the same country; and they have been referred to species ranging in size from *D. Marsupialis* to the minute *D. pusilla.*

In the Tertiaries of Argentina teeth and jaws of Opossums have been discovered which are assigned to extinct species more or less closely allied to living South American ones; these remains being found as low down as the Santa Cruz beds of Patagonia, which probably belong to the Miocene division of the Tertiary period.

The Miocene rocks of the United States have likewise yielded extinct species of Opossums, which may probably be assigned to the existing genus; unless, indeed, as has been stated to be the case, they differ by the absence of any inflection of the angle of the lower jaw.

Still more numerous are the Opossum-remains from the Upper Eocene, or Lower Oligocene, Tertiary beds of Europe, which have been assigned to a very large number of species. Although these extinct Opossums have been very generally separated under the name of *Peratherium*, there can be little doubt that they really belong to the existing genus *Didelphys*, which is consequently one of great antiquity. It should be mentioned, however, that we are still unacquainted with the number of incisor teeth in these Tertiary European Opossums. Remains of the genus have been obtained from the Tertiary beds of Hordwell in Hampshire, from the equivalent deposits of Débruge in Vaucluse, and of Montmartre near Paris, and likewise from the so-called Phosphorites of Quercy in the South of France.

It appears accordingly that during the middle portion of the Tertiary period Opossums were widely spread over the Northern Hemisphere, where they probably took origin from the extinct generic types of Marsupials of the antecedent Cretaceous epoch. With the development of the higher forms of Mammalian life they disappeared from Europe before the upper part of the Miocene, or middle division of the Tertiary period, and gradually retreated to a great extent from the northern half of the New World to find a secure refuge in South America. It is important to notice that there are no other Tertiary Marsupials in either Europe or North America, thus justifying the conclusions arrived at above as to a former connection between South America and Australia.

An interesting anecdote in connection with the demonstration of the affinities of *Didelphys gypsorum* of the gypsum quarries of Montmartre, as detailed by Owen, may be appropriately introduced here. The affinities of this fossil were originally deduced from the characters of the jaws and teeth; "but these were associated with other parts of the skeleton in the same block of stone." When Cuvier expressed his convictions of the Opossum nature of the fossil from the parts first examined, his scientific associates were incredulous. He invited them, therefore, to witness a crucial test. On the slab containing the jaws and teeth, the outline of the back part of the pelvis was also exposed, the fore part being buried in the matrix. By his delicate use of the graving-tool, Cuvier brought to light that part with the two Marsupial bones in their natural position. He thus demonstrated that there had been buried in the soft fresh-water deposits, hardened in after ages into the building-stone of Paris, an animal whose genus at the present day is peculiar to America.

FAMILY TRICONODONTIDÆ.

The whole of the remaining families of the Marsupials are mainly confined to the Secondary rocks, although a few survived into the earliest portion of the Tertiary period. All are of small size, and some are extremely minute.

In the present family the molar teeth, as shown in Fig 3 of the accompanying diagram, consist of three simple compressed and cutting cusps arranged in the same longitudinal line; the upper teeth biting on the outer side of the lower ones. In the upper jaw the number of teeth is unknown, but in the lower jaw there were three pairs of incisors, one of canines, four of premolars, and either three or four molars. As in existing Marsupials, but a single pair of teeth in each jaw was replaced by a vertical successor, and the angle of the lower jaw was inflected. The canine teeth were emplanted by two distinct roots ;—a character to which there is an approximation in some of the Bandicoots, where the single root of these teeth is partially divided by a deep vertical groove.

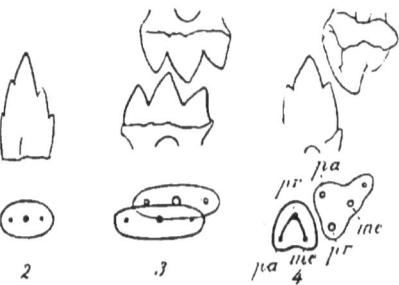

Diagram of Molar Teeth of Secondary Marsupials, seen from the outer side, and superposed in mutual relation. 2 *Dromatherium*, 3 *Triconoaon*, 4 *Spalacotherium*. (From Osborn.)

GENUS TRICONODON.

Triconodon, Owen, Encyclopædia Britannica, 8th ed., vol. xvii., p. 161 (1859); Lydekker, Cat. Foss. Marsup. Brit. Mus., pt. v., p. 258 (1887).

This, the single well-established genus of the family, is represented by three species from the Purbeck beds of Dorsetshire, belonging to the upper portion of the Jurassic system, of which the largest (*T. major*) may be compared in size to the existing *Dasyurus viverrinus*. Allied forms from the Upper Jurassic rocks of the United States have been assigned to a distinct genus, although they do not really appear separate from the present one.

FAMILY SPALACOTHERIIDÆ.

GENUS SPALACOTHERIUM.

Spalacotherium, Owen, Mesozoic Mamm., p. 53 (1871); Lydekker, Cat. Foss. Mamm. Brit. Mus., pt. v., p. 292 (1887).

In this family and genus the cusps or cones of the molar teeth (as shown in Fig. 4 of the diagram on page 274) are arranged in a triangle, of which the apex points inwards in the upper, and outwards in the lower jaw. In the lower jaw there are only three such cusps to each tooth, but in the upper teeth there is an additional posterior cusp. In structure these teeth are similar to those of the Marsupial Mole. All the members of the family are exceedingly minute, the lower jaw of the typical *S. tricuspidens*, of the Dorsetshire Purbeck beds, scarcely exceeding an inch in length. The number of teeth in the lower jaw was large, there being three pairs of incisors, one of canines, four of premolars, and six of molars. Typically occurring in the Dorsetshire Purbeck rocks, the family is represented by closely allied forms in the Upper Jurassic rocks of the United States.

FAMILY AMPHITHERIIDÆ.

This extensive family is taken to include a number of forms, of which it is not easy to give a collective definition, although they appear to differ from the two preceding families in having four pairs of incisor teeth, like the Opossums. The lower molars never consist solely of three simple cusps arranged in a straight line like those of the *Triconodontidæ*, or in a triangle like those of the *Spalacotheriidæ*.

GENUS PHASCOLOTHERIUM.

Phascolotherium, Owen, Trans. Geol. Soc., s.r. 2, vol. i., pt. i, p. 58 (1841); Lydekker, Cat. Foss. Mamm., pt. v., p. 270 (1887).

This genus, as represented by *P. bucklandi* of the lower Jurassic rocks of Stonesfield, near Oxford, was one of the earliest discovered of the Secondary Marsupials, and is characterised by having only seven pairs of cheek-teeth (including premolars and molars). The molars had three cusps arranged in a longitudinal line, of which the middle one was much larger than the others, while there were minute accessory cusps at the two extremities of each tooth, and a distinct ledge at the base of the inner side. The whole length of the lower jaw was only about an inch.

GENUS AMPHILESTES.

Amphilestes, Owen, Encyclopædia Britannica, 8th ed., vol. xvii., p. 157 (1859); Lydekker, Cat. Foss. Mamm. Brit. Mus., pt. v., p. 271 (1887).

In this genus, as represented by *A. broderipi*, of the Stonesfield rocks, the molar teeth were of the same general structure as in *Phascolotherium*, but much more numerous, the total number of teeth being probably the same as in the next genus. The one species was exceedingly minute.

GENUS AMBLOTHERIUM.

Amblotherium, Owen, Mesozoic Mamm., p. 29 (1871); Lydekker, Cat. Foss. Mamm. Brit. Mus., pt. v., p. 274 (1887).

A second group of the family is represented by the minute *Amblotherium soricinum*, of the upper Jurassic Purbeck rocks of Dorsetshire, which is selected from among several allied genera on account of the number of lower teeth being fully known. In this genus there were four pairs of incisor, one of canine, four of premolar, and either seven or eight of molar teeth. The latter differ from those of the preceding genera, and thereby resemble the corresponding lower teeth of the existing Opossums and Bandicoots, in that they consist of an anterior portion carrying three cusps arranged in a triangle, and of a posterior moiety, or heel. It has only quite recently been ascertained that the earliest known of all the Secondary Marsupials, namely, the *Amphitherium* of Stonesfield, which was described by the French naturalist De Blainville as far back as the year 1838, had teeth of this type.

The great interest attaching to *Amphilestes*, *Amblotherium*, *Amphitherium*, and their allies, is that they, and they alone among Mammals, had molar teeth comparable in form, and to a certain extent in structure, with those of the existing Banded Ant-eater (*Myrmecobius*) of Australia, which, as already mentioned, may be regarded as the sole and specially modified descendant of these ancient forms of Mammalian life.

Several genera allied to *Amblotherium* are represented in the upper Jurassic rocks of North America; amongst them one which has received the name of *Dryolestes* may be specially mentioned; and it appears that in that continent certain members of the families survived till the succeeding Cretaceous epoch.

FAMILY DROMATHERIIDÆ.

GENUS DROMATHERIUM.

Dromatherium, Emmons, American Geology, pt. vi., p. 93, 1857.

The earliest, and at the same time the most generalised

Mammals that can by any possibility be included in the Marsupials, are two minute ones from the reputed Triassic rocks of North Carolina, one of which has received the name of *Dromatherium sylvestre*. The total number of teeth in the lower jaw is fourteen on each side, three of which are incisors, one a canine, three premolars, and the remaining seven molars. The latter differ from those of all Mammals in which the teeth are differentiated into series, in having imperfectly divided roots, thus approximating to those of reptiles. The premolars are nearly simple cones; while the molars, as shown in Fig. 2 of the diagram on page 274, consist of one main cone, with minute accessory cusps on the edges; and it is probable that the opposing teeth of the upper and lower jaws mutually interlocked, thus displaying another reptilian feature.

Whether this exceedingly primitive creature was really a Marsupial may be open to doubt, even if it be a Mammal at all. Assuming, however, that, as is probable, it is really Mammalian, it may equally well indicate an Order more nearly allied to the Monotremes than to the Marsupials, being, in fact, an ancestral type of the former. Our information is, however, at present far too meagre to admit of anything definite being said on this point.

ORDER MONOTREMATA.

As already mentioned, our knowledge of the past history of this group is practically a blank; the only fossil forms that can be definitely included in it being an extinct representative of each of the existing families.

FAMILY ORNITHORHYNCHIDÆ (*suprà*, p. 230).
GENUS ORNITHORHYNCHUS (*suprà*, p. 231).
1. ORNITHORHYNCHUS AGILIS.

Ornithorhynchus agilis, De Vis, Proc. Roy. Soc., Queensland, 1885, p. 35.

This species is described on the evidence of one ramus of the lower jaw, and a tibia, or bone of the lower leg, which apparently belonged to an adult animal, and thus indicate that the fossil Duck-bill was a considerably smaller animal than its existing representative.

Distribution.—Queensland.

FAMILY ECHIDNIDÆ (*suprà*, p. 240).
GENUS ECHIDNA (*suprà*, p. 241).
1. ECHIDNA OWENI.

Echidna oweni, Krefft, Ann. Mag. Nat. Hist. ser. 4, vol. i., p. 113 (1863); Lydekker, Cat. Foss. Mamm. Brit. Mus., pt. v., p. 295 (1887).

Echidna ramsayi, Owen, Phil. Trans., 1884, p. 273.

A species considerably exceeding in size the existing Papuan *Proechidna bruijni*, and possibly referable to the same genus.

Distribution.—New South Wales.

ORDER MULTITUBERCULATA.

As already mentioned, the serial position of that remarkable group of extinct Secondary Mammals, generally known by the name of Multituberculata, cannot at present be definitely determined, although the balance of evidence is in favour of their being more or less intimately related to the existing Monotremes. Whether, however, they should be included in the same order with the latter, or whether, as is more probable, they should constitute an order by themselves, is likewise a matter of uncertainty. If they really indicate a distinct order, it may prove that the Monotremata and Multituberculata constitute two ordinal groups in the sub-class Prototheria.

From a superficial resemblance between the teeth of certain members of the family and those of the Rat-Kangaroos, it was at one time considered that the Multituberculata belonged to the Diprotodont Marsupials, but subsequent investigations have led to the abandonment of this view.

The main reasons for placing this group in the neighbourhood of the Monotremes are, firstly, that while their molar teeth are quite unlike those of all other Mammals, they present a certain distant resemblance to the deciduous teeth of the Duckbill; and, secondly, that in the one instance where the bones of the shoulder-girdle have been discovered, these are of the characteristic Monotrematous type; that is to say there is a distinct coracoid and metacoroid, and probably, therefore, an interclavicle.

Ranging throughout the whole of the Secondary period, that is to say from the Triassic to the Cretaceous rocks inclusive, the Multituberculata also survived into the lowest division of the Tertiary epoch. Geographically their distribution was extensive, embracing Europe, North America, and South Africa. Whether they occur in South America is, however, doubtful, as it is by no means certain whether the forms from the Tertiary deposits of that continent which have been assigned to the group are rightly referred.

One of the most remarkable features connected with these ancient Mammals is the specialised nature of their dentition; a specialisation so great that it is quite certain they could not

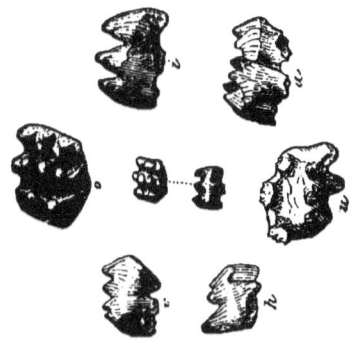

Six views of an Upper Molar Tooth of a Multituberculate Mammal. The two small figures in the centre of the group indicate the natural size of the specimen.

have been the ancestors of Mammals in general; although this would be no bar to their having given origin to the existing representatives of the Monotremes. If it should prove that the Multituberculata have any ancestral relationship to the latter, it would indeed go far towards the confirmation of the suggestion that the Monotremes are but remotely allied to the higher Mammals, even if they have a right to the title of Mammals at all.

As a group, the Multituberculata are characterised by their

molar teeth carrying two or three longitudinal ridges supporting numerous small tubercles, such ridges being separated by deep grooves, as shown in the accompanying figures, and the number of ridges in the upper molars being usually one in excess of those of the corresponding lower teeth. As in the Diprotodont Marsupials, there is a single pair of large and sharp incisor teeth in the lower jaw; and a further superficial resemblance to that group is often shown by the last premolar being modified into a large compressed and cutting tooth with well-defined parallel groovings and ridges on the sides of its crown.

The numerous genera of the Multituberculata are arranged under several families, which will form the headings under which our brief remarks will be arranged.

FAMILY TRITYLODONTIDÆ.

GENUS TRITYLODON.

Tritylodon, Owen, Quart. Journ. Geol. Soc., vol. xl., p. 146 (1884); Lydekker, Cat. Foss. Mamm. Brit. Mus., pt. v., p. 201 (1887).

In this family the upper molar teeth have three longitudinal ridges, as shown in the last illustration, and are broader than long; while the premolars are of the same general structure, but somewhat simpler. In the upper jaw there is one pair of large, somewhat chisel-like incisors, followed by a second very small and functional pair; then, owing to the absence of a canine, comes a very long gap, followed by the two premolars, and these again, by the four molars. The lower jaw is unknown.

This genus is typically represented by *T. longævus*, from the Secondary rocks of South Africa, which was an animal of the approximate size of a Rabbit. A smaller species, of which an upper molar tooth is represented in the figure on page 281, has been obtained from the Triassic rocks near Strasbourg.

FAMILY BOLODONTIDÆ.
GENUS BOLODON.

Bolodon, Owen, British Mesozoic Mamm., p. 6 (1871); Lydekker, Cat. Foss. Mamm. Brit. Mus., pt. v., p. 203 (1887).

Nearly allied to the preceding family is a second one, typically represented by the genus *Bolodon*, first described from the upper Jurassic Purbeck rocks of Dorsetshire, but apparently also occurring in the corresponding beds of the United States, where, however, it has received a distinct name.

In this family the four upper molars are longer than broad, and each carry only two longitudinal ridges, separated by a deep median groove, and each bearing several blunt tubercles. The upper premolars have triangular crowns, surmounted by three tubercles.

Minute molar teeth very similar to those of *Bolodon* from the Triassic rocks of Stuttgart and Somersetshire, described by the name of *Microlestes*, are perhaps referable to the present family, although in the absence of any knowledge as to the nature of the premolars, it is by no means certain that the genus should not rather be assigned to the undermentioned *Plagiaulacidæ*.

FAMILY POLYMASTODONTIDÆ.
GENUS POLYMASTODON.

Polymastodon, Cope, American Naturalist, vol. xvi., p. 684 (1882); Lydekker, Cat. Foss. Mamm. Brit. Mus., pt. v., p. 200 (1887).

The lower Eocene rocks of New Mexico have yielded the

remains of a large member of the present group, which appears entitled to constitute a family by itself. Its leading characters are to be found in the reduction of the molar teeth to two, and of the premolars to a single pair in each jaw. The molars are much longer than wide, those of the upper jaw having three, and those of the lower jaw two, longitudinal ridges; each ridge having numerous broad blunt tubercles, and the longitudinal grooves so narrow as to be but little wider than the secondary grooves between the tubercles. The single premolar, although simpler, is of the same general structure as the molars. It is noteworthy that the angle of the lower jaw is inflected in the same manner as in the Marsupials.

Polymastodon is one of the largest representatives of the Multituberculata, the length of the lower jaw being about four inches.

FAMILY PLAGIALAUCIDÆ.

GENUS PLAGIAULAX.

Plagiaulax, Falconer, Quart. Journ. Geol. Soc., vol. xiii., p. 261 (1857); Lydekker, Cat. Foss. Mamm. Brit. Mus., pt. v., p. 196 (1887).

This last well-defined family of the group under consideration is typically represented by the genus *Plagiaulax* from the upper Jurassic rocks of Dorsetshire, but likewise includes other genera from the Cretaceous and lower Tertiary rocks of the United States, and yet others from the lowest Tertiary strata of North America and France.

Agreeing with the preceding family in the elongated form of the molar teeth, which likewise carry three longitudinal ridges in the upper, and two in the lower jaw, the *Plagiaulacidæ* are specially characterised by the peculiar conformation of the premolar teeth, which may vary in number from one to four pairs; the molars being always two. These peculiar premolars, as shown in the accompanying cut, are of relatively large size, and are characterised, when unworn, by having a highly compressed, convex crown, with its summit narrowed to a sharp cutting edge, and its sides usually marked by oblique ridges and grooves. The single pair of lower incisor teeth are curved and pointed.

It was from the resemblance of their remarkable premolar teeth to the last premolar tooth of the Rat-Kangaroos, coupled with the presence in both groups of a single pair of large lower incisors, that led to the Multituberculata being included among the Diprotodont Marsupials. An important point of

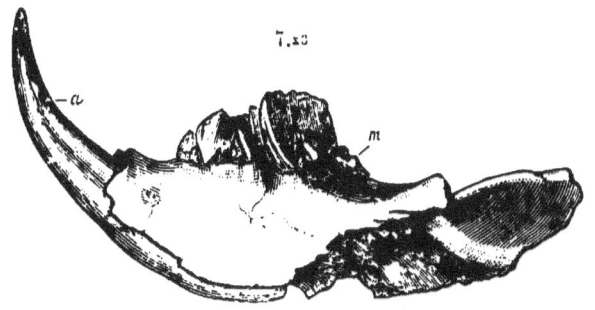

One half of an imperfect Lower Jaw of a North American Representative of the *Plagiaulacidæ*. Much enlarged. (After Marsh.)

difference between the two groups is, however, to be found in the circumstance that whereas in the Rat-Kangaroo the summit of the crown of the last premolar is slightly concave, in all the *Plagiaulacidæ* it is markedly convex.

It is noteworthy that the earlier members of the family have a large number of premolars, and but few grooves on the last tooth of that series, whereas in the later forms the number of premolars diminishes, while the grooves on the last premolar become more numerous.

The nature of the food of *Plagiaulax*, all the species of

which are of minute size, gave rise to a bitter conflict analogous to the one which, as we have seen, occurred at a later date in the case of the Marsupial *Thylacoleo;* one party maintaining that these little creatures were carnivorous, while the other as persistently asserted their herbivorous nature. The evidence in favour of the latter view was mainly furnished from the similarity of the teeth of these extinct Mammals to those of the Rat-Kangaroos, and although this similarity is now regarded merely as an instance of parallel development, and not as indicative of genetic affinity, it is probable that the deductions drawn therefrom are true, and that these early British Mammals were as herbivorous as the antipodean group they so curiously simulate in the structure of their teeth.

APPENDIX.

This appendix contains such forms as have been named since the work was first printed, together with those that have been recognised as entitled to distinction since the same date.

DIPROTODONTS.

SUB-ORDER DIPROTODONTIA (*supra*, p. 11).
FAMILY MACROPODIDÆ (*supra*, p. 11).
GENUS DENDROLAGUS (*supra*, p. 58).

V. BENNETT'S TREE-KANGAROO. DENDROLAGUS BENNETTIANUS.

Dendrolagus bennettianus, De Vis, Abstracts Proc. Linn. Soc. N. South Wales, 1886, p. 5; Sclater, Proc. Zool. Soc., 1894, p. 693, pl. xlvi.; Waite, Proc. Linn. Soc. N. South Wales, ser. 2, vol. ix., p. 571 (1894).

The first description of this form was so insufficient that it was not definitely recognised by Thomas in his Cat. Marsup. Brit. Mus. (p. 96) as entitled to rank as a valid species.

Characters.—Fur on the back directed backwards; colour of body mouse-brown; head and sides of neck rufescent; patch on back above the tail black; feet blackish; tail black inferiorly and at the tip, lighter on upper surface. Length of head and body about 24 inches; of tail about 30 inches. The hinder part of the palate is peculiar in having a pair of small unossified vacuities.

Mr. Waite remarks that externally this species appears to most nearly resemble *D. inustus*. The skull also agrees with that species and with *D. ursinus* in the non-inflated forehead, but differs from these and agrees with *D. lumholtzi* in the formation of the fronto-nasal suture.

Distribution. Queensland.

Habits.—Mr D. Le Souef (Victorian Naturalist, vol. xi., p. 11, 1894) writes that "the Tree-climbing Kangaroo (*Dendrolagus bennettianus*) is generally found on or near the top of the ranges, where the timber is not so high or difficult to climb. They remain during the day on the highest branches of a tree, and descend at night to pass from one tree to another. They seem to feed on birds' nest ferns, leaves of certain trees, creepers, and probably on wild fruits."

FAMILY PHALANGERIDÆ (*supra*, p. 75).
GENUS PHALANGER (*supra*, p. 86).

IIIa. SHORT-HEADED CUSCUS. PHALANGER BREVICEPS.

Phalanger orientalis, var. *breviceps*, Thomas, Cat. Marsup. Brit. Mus., p. 204 (1888).

Phalanger breviceps, Thomas, Novitat. Zool. ii., p. 165 (1895).

This form is now recognised by its describer as a distinct species.

Habitat.—New Britain Group, and Solomon Islands as far east as San Christoval.

GENUS PSEUDOCHIRUS (*supra*, p. 92).

XI. DAHL'S RING-TAILED PHALANGER. PSEUDOCHIRUS DAHLI.

Pseudochirus dahli, Collett, Zool. Anzeiger, 1895, p. 464.

Characters—Size large; fur long and woolly; general colour reddish-grey above, with middle line of forehead blackish; tail more rufous, without white tip; chest-spot rufous. Head small; tail very short, about half the length of the body, with the tip almost naked; ears short; muzzle very narrow. Orbital ridges of skull parallel, and not uniting behind; large vacuities in hinder part of palate. Incisors and molars stout; intermediate teeth feebly developed or wanting; second upper incisor elongated horizontally; first lower incisor lancet-shaped. Length to root of tail about 17·5 inches ; of tail 9·5 inches. (*Collett*).

Distribution.—Mary river, North Australia.

Habits.—This species, which in general colour is very similar to *P. peregrinus*, is peculiar in dwelling among rocks, instead of on trees. During the daytime it conceals itself amongst huge granite rocks, which it leaves at night to ascend trees in search of food. It feeds chiefly on the soft parts of a berry with a large stone, somewhat like a cherry, but larger. It never resorts to hollow trunks for sleeping, although, when hunted, it will occasionally take refuge in a tree.

FAMILY EPANORTHIDÆ.

This family was originally made known to science upon the evidence of fossilized remains from the middle Tertiary beds of Santa Cruz, in Patagonia, where the allied extinct families *Abderitidæ* and *Garzoniidæ* are likewise met with; and it was not till the autumn of 1895 that a living representative of the group was recognised. The *Epanorthidæ*, as typified by the extinct genus *Epanorthus*, differ from all other existing Diprotodonts in the absence of syndactylism in the hind foot; and it is probable that the same condition obtained in the other two extinct families. The family may be defined as non-syndactylous Diprotodonts, with a hallux present, which is not widely opposable, and four upper and three lower pairs of incisor teeth, the last lower premolar being of ordinary form and dimensions. It is obvious that the inclusion of this family in the Diprotodontia will render necessary some modification of the definition given on page 11. Possibly, however, the present group may really form a sub-order by itself, in which event the name Paucituverculata, proposed by Señor Ameghino, is available.

THE SELVAS. GENUS CŒNOLESTES.

Hyracodon, Tomes, Proc. Zool. Soc., 1863, p. 50, *nec* Leidy, 1856.
Cœnolestes, Thomas, Ann. Mag. Nat. Hist., ser. 6, vol. xvi, p. 367 (1895); Proc. Zool. Soc., 1895, p. 870

Although a member of the genus has been known since 1863, its true zoological position was not recognised till the discovery of a second form in 1895.

The generic characters, as given by Mr. Thomas, are as follows: General appearance not unlike that of a rat or small opossum. External characters very much as in the Dasyurid genus *Phascologale*. Head elongate; nose naked, both in front and on the top of the muzzle; ears short, squarish, their inner surface provided with several tragus-like projections. Fore-feet with five toes, of which the outer one, as well as the thumb, has a distinct nail, while each of the three middle digits bears a curved claw. The third digit is the longest, the second and third being sub-equal and slightly shorter; the fifth reaching to the end of the first joint of the fourth, and the first to the middle of the corresponding segment of the second. Soles (palms) of fore-feet naked, with one elongated wrist-pad, three at the base of the toes, and one on the thumb. Hind-feet of normal shape, neither syndactylous nor—as in the opossums—modified into a hand. Great toe (hallux) short, clawless, and not thoroughly opposable; other toes clawed and sub-equal, the fourth being slightly the longest; soles naked, with six pads of somewhat elongated form, and not transversely striated. Tail long, slender, rat-like, and so thinly haired as to appear naked, the terminal inch of the lower surface being completely so; probably prehensile. A small rudimental pouch in the female.

As regards the skull, the founder of the genus observes that in general proportions this is "something like that of a *Perameles*, although thinner and more delicately built, with a similarly elongated muzzle, smooth and rounded brain-case, and obsolete supraorbital and cranial crest and ridges; the zygomata are, however, so much more boldly expanded as somewhat to spoil the resemblance, which in any case does not apply to details. Nasals long, thin, anterior two-thirds narrow, almost parallel-sided, but a little tapering forwards, their posterior third well expanded, somewhat as in ordinary *Didelphys*, but not expanded enough to meet the upper edge of the maxilary bone. As a result, an ante-orbital vacuity is left on each side in the position of, and formed exactly in the same way as, that of so many Ruminants. Apart from the latter group, this vacuity is perfectly unique among Mammals, and therefore well worthy of special note."

The dentition may be expressed by the formula i. $\frac{4}{3}$; c. $\frac{1}{1}$; p. $\frac{3}{3}$; m. $\frac{4}{4}$; = 46. Generally, the teeth present a considerable resemblance to those of the Australian *Dromicia*, especially as regards their relative proportions. The canines are as well developed as in an ordinary carnivorous Marsupial. By Mr. Thomas the molars are described as "low-crowned, with low, rounded, or scarcely pointed cusps, not unlike those of *Petaurus* or *Dromicia;* the two anterior square, quadricuspidate, although apparently there are only three roots to each, the postero-internal cusp being placed on a sort of flange overhanging the palate, and not supported by a root; third molar similar, but without the extra postero-internal cusp; last molar minute, triangular, as small in cross section as the last incisor."

The selvas are so called on account of the type specimen of the larger species having been discovered on an estate of that name. Their special interest centres on the evidence they, in association with the occurrence of fossil Dasyurine Marsupials in Patagonia, afford of a former land connection between Australia and South America, to which allusion has already been made on page 5 of the text.

The genus may be defined as follows: Form as in *Phascologale*; fifth fore-digit with a nail; tail long and more or less prehensile; a rudimental pouch. In the skull large preorbital vacuities are present; the palate is very imperfectly ossified; and there are three pairs of premolar teeth in each jaw, of which the two hinder lower pairs are large and functional.

1. ECUADOR SELVA. CŒNOLESTES FULIGINOSUS.

Hyracodon fuliginosus, Tomes, Proc. Zool. Soc., 1863, p. 50, pl. viii.

Cœnolestes fuliginosus, Thomas, Proc. Zool., 1895, p. 877.

Characters.—Apparently mainly distinguished from the second species by its considerably inferior size, which may be roughly compared to that of a water-shrew.

Distribution.—Ecuador.

II. BOGOTA SELVA. CŒNOLESTES OBSCURUS.

Cœnolestes obscurus, Thomas, Ann. Mag. Nat. Hist., ser. 6, vol. xvi., p. 367 (1895); Proc. Zool. Soc., 1895, p. 877, pl. ii.

Characters.—Size approximately equal to that of a small rat; fur soft, thick and close; general colour uniform bistre-brown, becoming rather darker along the middle line of the back, and very slightly paler below. Ears short, brown, and almost completely naked; feet brown; tail about as long as head and body, slender, very finely haired, with the terminal portion naked inferiorly.

Distribution.—Bogota, Colombia.

Habits.—All that is known of the habits of this interesting little Marsupial is comprised in a letter from Mr. G. D. Child to Mr. Thomas. The former gentleman writes that "the little animal you speak of is called the 'Raton Runcho,' which means 'Opossum Rat.' It lives in the high brushwood, and is supposed to feed on birds' eggs and small birds. It is very rare indeed, and is obtai ed wi h much difficu'ty." This indicates that the creature belongs to a fast-waning type. In general habits it would appear to be very similar to some of the Phalangers, with which its resemblances are the closest.

POLYPROTODONTS.

SUB-ORDER POLYPROTODONTIA (*suprà*, p. 128).

FAMILY DASYURIDÆ (*suprà*, p. 150).

GENUS PHASCOLOGALE (*suprà*, p. 166).

I. CREST-TAILED POUCHED MOUSE.   PHASCOLOGALE CRISTICAUDA.

Phascologale cristicauda (*suprà*, p. 169); Spencer, Rep. Horn Expedition—Zool., p. 21 (1896).

Characters.—Redescribed by Mr. Spencer as follows: "Size large; form strong; fur close and soft, mainly composed of the under-fur. The general body-colour is mouse-grey, ting d with rufous on the back. The under surface is white or cream-coloured, and so are the inner and anterior faces of the limbs and the upper surfaces of the hands and feet. The under-fur on the back is slate-grey at the base, and rufous terminally; on the ventral surface it is cream-white terminally. The tail is thickly covered in its proximal (basal) half on the upper and lateral surfaces with coarse, chestnut-coloured hair; on the ventral aspect the hairs are dark brown in colour. About the middle of its length it is covered with coarse black hairs, which increase in length distally on the upper and under surface until—especially on the upper surface—they form a distinct black crest, a smaller crest being present on the under surface. The tail is considerably swollen out proximally, and somewhat thickened, although the thickening is hidden by the body-hairs, which pass on to the root of the tail. The ears, when laid forward, reach to the posterior border of the eye. They are covered externally and internally with short, stiff hair. The eye is surrounded by a light ring of hair. The hairs on the fold of the pouch and the pouch-area—where they are scanty—are white. Hands and feet white, or light grey above; palms with six granulated elevations, each with a small unstriated pad. There is a small tuft of white, whisker-like hairs on the posterior side of the fore-arm just above the wrist. Soles of feet with three granulated elevations at the base of the toes, each with a small unstriated pad. The soles are hairy in the heel region, and have a series of thick-set strong hairs running along the outer and inner margins and bending over on to the under surface, only the medium part of which, so far back as the heel, is really naked. The medium part is strongly granulated. Pouch opening vertically downwards, with moderately developed lateral folds. Mammæ eight (may be reduced to six, or, rarely, four)."

Distribution.—South and Central Australia.

XIII. LESSER BRUSH-TAILED POUCHED MOUSE. PHASCOLOGALE CALURA.

Phascologale calura (*suprà*, p. 176); Spencer, Rep. Horn Expedition—Zool., p. 30 (1836).

Distribution.—Recorded by Spencer from Central Australia.

XIV. MACDONELL'S POUCHED MOUSE. PHASCOLOGALE MACDONELLENSIS.

Phascologale macdonellensis, Spencer, Proc. Roy. Soc. Victoria, ser. 2, vol. vii., p. 222 (1895); Rep. Horn Expedition—Zool., p. 27, pl. ii., fig. 1, pl. iv., figs. 9—12 (1896).

Characters.—Size medium; fur moderately coarse; general colour of back dull greyish brown, with a chestnut patch behind the ear; under surface grey; a light line above and below eye; a light line along upper jaw, meeting light line of under surface behind angle of same. Ears rounded, clothed on both sides with short, light hairs, reaching, when laid forward, at least as far as middle of eye. Fore- and hind-feet grey above; soles of fore-feet with six striated pads; a small tuft of white whisker-like hairs on hinder side of fore-arm above the wrist; soles of hind-feet naked, except under the heel, where they are hairy, granulated; the pad on the great toe (hallux) double; all six pads distinctly granulated. Tail shorter than head and body; stout for its basal half, tapering rapidly about the middle, and thence rapidly to tip; markedly thickened; covered with longish stiff hairs, but devoid of crest; in colour somewhat lighter than body. Pouch slightly developed, opening downwards, with two lateral folds of skin. Teats six, in three pairs. Last upper primolar minute, and corresponding lower tooth absent. (*Spencer*).

Distribution.—Central Australia.

Habits.—The animal is terrestrial, living in holes, amongst rocks, and under stones.

GENUS SMINTHOPSIS (*suprà*, p. 177).

V. LARAPINTA POUCHED MOUSE.   SMINTHOPSIS LARAPINTA.

Sminthopsis larapinta, Spencer, Proc. Roy. Soc. Victoria, ser. 2, vol. viii., p. 8 (1896); Rep. Horn Expedition—Zool., p. 33, pl. ii., fig. 2 (1896).

Characters.—Size small; form light and delicate; fur very soft and fine, moderately long, composed almost entirely of underfur, with a few longer dark hairs. General colour mouse-grey, suffused on back with rufous; sides, under-parts, and upper surfaces of both pairs of feet white. Ears large, when laid forward reaching considerably beyond the eye. Soles of fore-feet naked, granulated, with the postero-external pad V-shaped and striated, its apex directed forwards. A small tuft of white whisker-like hairs on hinder aspect of fore-arm above wrist. Soles of hind feet granulated in front and middle; four pads, one at base of great toe, and three larger ones at bases of other toes, the latter faintly striated. Tail much longer than body; very stout basally, and thickened, tapering to a whip-like point; strongly scaled, with hairs at base so short as n't to hide scales. The characteristics of this species are the long, stout, and much thickened tail, and the great relative length and width of the hind-foot. (*Spencer*).

Distribution.—Central Australia.

Habits.—Frequents the stony table-lands, whereas *S. crassicaudata* prefer softer low ground near the creeks and amongst the sand hills.

VI. SAND-HILL POUCHED MOUSE. SMINTHOPSIS PSAMMOPHILUS.

Sminthopsis psammophilus, Spencer, Proc. Roy. Soc. Victoria, ser. 2, vol. vii., p. 223 (1895); Rep. Horn Expedition—Zool., p. 35, pl. i., fig 2 (1896).

Characters.—Size medium; fur close, long, and fine, with some longer, stiff, and darker hairs interspersed on the back. General colour of upper-parts dark grey, of under-parts white; a brownish tinge on the thighs; hairs of back grey at the roots, but darker at the tips; those of under surface grey inferiorly and white superiorly; a ring of white hairs round the eye, and a small tuft of white bristly hairs above the wrist; hair on both pairs of feet cream-coloured. Ears covered on both aspects with short, stiff, grey hairs; when turned forwards, reaching half-way between the eye and the muzzle. Soles of fore-feet granulated, with six ill-defined elevations, but devoid of striated pads; those of hind-feet hairy, with three non-striated pads placed on granular elevations at the bases of the toes; great toe (hallux) relatively small, and situated about midway between the heel and the bases of the claws of the other digits. Tail long, thin, clothed with short, stiff, whitish hairs above and on the sides, and below with a line of black hairs. (*Spencer*).

Distribution.—Central Australia.

Habits.—Diurnal and terrestrial, living among sand-hills covered with porcupine-grass (*Triodia irritans*).

GENUS DASYUROIDES.

Dasyuroides, Spencer, Proc. Roy. Soc. Victoria, ser. 2, vol. viii., p. 5 (1896); Rep. Horn Expedition—Zool., p. 36 (1896).

Size small in comparison with *Dasyurus*; general build comparatively stout; tail long; hallux absent. Feet long and strong, not delicate as in *Sminthopsis*; toes with strong, sharp, curved claws; soles of both fore- and hind-feet very hairy, with the median part granulated; those of the hinder pair with three well-marked pads placed on granulated elevations at the base of the toes.

Its describer remarks that the type of this genus cannot be placed in either *Phascologale* or *Sminthopsis*, as at present defined. " In certain respects it presents characters at present regarded as distinctive of one or the other, while it differs markedly from both in the absence of a hallux. To have associated it with these forms would have meant the merging of the two genera in one another, and the additional widening of the characters so as to include a non-hallucated form. The only other alternative was the creation of a new genus. and I therefore adopted this plan *Dasyuroides* may be regarded as a genus closely allied to both *Phascologale* and *Sminthopsis*, and serving at the same time as an approach to *Dasyurus*. The general form of the body closely resembles that of the larger Phascologales or of a very small *Dasyurus*, and is very different from—that is, much less slender than—that of even the largest *Sminthopsis*."

BYRNE'S POUCHED MOUSE. DASYUROIDES BYRNEI.

Dasyuroides byrnei, Spencer, Proc. Roy. Soc. Victoria, ser. 2, vol. viii., p. 6 (1896); Rep. Horn Expedition—Zool., p. 36, pl. iii., pl. iv., figs. 1–4 (1896).

Characters.—Size and appearance very similar to that of *Phascologale cristicauda*, from which it may be distinguished by the absence of the great toe (hallux). Fur soft and close, composed mainly of under-fur; general colour grizzled grey, with a faint rufous tinge, especially on the head and back; underparts and upper surfaces of fore- and hind-feet white. Tail rufous for rather less than its basal half, where it is thickly covered with longish hair; terminal half clothed with long black hairs, forming a crest above and beneath. A small tuft of bristly white hairs above the wrist, one or two of which are so elongated as to resemble the face-whiskers. Ears naked above, when laid forwards, reaching nearly to the front angle of the eye. Soles of fore-feet with distinct, faintly striated pads placed on granular elevations; those of hind-feet narrow, with three well-marked pads situated on elevations at the bases of the toes, and distinctly striated; middle region of sole naked and granulated; each side with a close series of hairs inclining to the middle line. Tail fairly stout, but not specially thickened. Teats six; pouch rudimental, with two low lateral folds. Dentition very similar to those species of *Phascologale* in which the last lower premolar is wanting. (*Spencer*).

Distribution.—Central Australia.

Habits.—Terrestrial, nocturnal, and insectivorous; living in burrows on sandy and stony table-lands.

APPENDIX.

FAMILY DIDELPHYIDÆ (*suprà*, p. 193).

GENUS DIDELPHYS (*suprà*, p. 196).

By Mr. Thomas (Ann. Mag. Nat. Hist., ser. 2, vol. xiv., p. 184, 1894) the sub-genera *Metachirus*, *Philander*, *Micoureus*, and *Peramys*, mentioned in the text, are admitted to generic rank. In a later memoir (*op. cit.*, vol. xviii., p. 31 , 1896) he adopts the name *Marmosa* in place of *Micoureus*, on account of priority.

I*a*. KOSERITZ'S OPOSSUM. DIDELPHYS KOSERITZI.

Didelphus koseritzi, Ihering, Os Mammiferos do Rio Grande d) Sul, p. 6 (1893).

Characters.—Allied to *D. marsupialis*, var. *aurita* and var. *azaræ*, which Ihering regards as probably entitled to rank as distinct species. Differs from the former by the uniformly blackish-brown head, which has a sub-circular ochrey spot behind the eye. Possibly identical with *D. marsupialis*, var. *azaræ*, which Ihering likewise thinks may form a distinct species.

Habitat.—Rio Grande do Sul.

V*a*. TRINIDAD OPOSSUM. DIDELPHYS TRINITATIS.

Didelphys (*Philander*) *trinitatis*, Thomas, Ann. Mag. Nat. Hist., ser. 6, vol. xiii., p. 438 (1894).

Characters.—The island form of *D. philander*, from which it differs by its inferior size, relatively smaller skull and limbs, but proportionately larger ears and tail. Fur close, soft, and somewhat straighter and less woolly than in the mainland form. Colour very similar to that of the latter, yellowish-rufous above and buff beneath; the face greyish white, with a narrow median dark line. Heel lacks the minute additional outer sole-pad generally found in *D. philander*. Tail very long, furred like the body for about its basal inch, this fur extending about a quarter of an inch further below than above; remainder of tail practically naked, but the lines between the scales clothed with microscopic hairs; in colour, uniformly brown above and slightly paler beneath, whereas in *D. philander* the entire tip is whitish. Teats seven. (*Thomas*).

Distribution.—Trinidad.

XIa. SHREW-LIKE OPOSSUM. DIDELPHYS SORICINA.

Didelphys soricina, Philippi, Arch. Naturgesch., 1894, vol. i., p. 36.

Characters.—Said to resemble "*D. parva*" (? = *pusilla*) in general characters, but altogether blackish above and whitish beneath. Ears large; tail somewhat longer than head and body, black above, white beneath, clothed above at the base with long hair, then sparsely haired, below, with the hairs posterior to the naked portion longer and more numerous.

Distribution.—S. Chili.

XII. GREY OPOSSUM. DIDELPHYS MARMOTA.

Didelphys marmota, Oken. Lehrb. Nat., vol. iii., p. 1140 (1816).
Didelphys grisea, Desmarest, Dict. Sci. Nat., vol. xlvii., p. 393 (1827); *suprà*, p. 212.
Micoureus griseus, Thomas, Ann. Mag. Nat. Hist., ser. 6, vol. xiv., p. 184 (1894).
Marmosa marmota, Thomas, *op. cit.*, vol. xviii., p. 314 (1896).

In the text the Brazilian *D. incana* was identified, on the authority of Thomas, with this species, which is confined to Paraguay and Corrientes. It is still known only by very few examples, and appears to be more nearly allied to the Chilian *D. elegans* than to *D. incana*, although externally the two are very different

Characters.—Size rather less than in *D. murina*; fur very short, soft, and fine; general colour of upper-parts very like that of a house-mouse; under-parts white, and sharply defined from dark of back by a broad band of paler grey. Eyes ringed with black, and the forward continuation of the pale lateral band causing the appearance of pale outer eye-rings, separated in the middle line by the darker middle line of the back. Ears large, leafy, and grey, with the anterior basal prominence little developed. Tail long and, except at the extreme base, naked; grey above for three-quarters its length, the terminal quarter and all the lower surface white. Front of fore arm paler grey than lateral band, upper surface of fore-foot white, as in the hind foot. (*Thomas*).

Distribution.—Paraguay and Corrientes.

XII*a*. HOARY OPOSSUM. DIDELPHYS INCANA.

Didelphys incana, Lund, Blik. Brazil, Dyr. Danks. Afhandl., vol. viii., p. 327 (1841); *supra*, p. 212.
Micoureus incanus, Thomas, Ann. Mag. Nat. Hist., ser. 6, vol. xiv., p. 186 (1894).
Marmosa incana, Thomas, *op. cit.*, vol. xviii., p. 313 (1896).

Characters.—Generally, those given in the text under the heading of *D. grisea*, but the grey of the upper-parts shading gradually into the white of the belly.

Distribution.—Central and Eastern Brazil.

XII*b*. MERIDA OPOSSUM. DIDELPHYS FUSCATA.

Marmosa fuscata, Thomas, Ann. Mag. Nat. Hist., ser. 6, vol. xviii., p. 313 (1896).

Characters.—Size rather smaller than in *D. incana*. Ears large, their internal basal projection small; a rounded lobe at the base of their outer edge. Fur close, soft, and velvety. General colour above a dark smoky or bistre-brown, a dull buffy tinge present on the fore back and sides. Under surface dirty whitish, the slaty bases of the hairs showing through; line of demarcation on sides fairly well defined. Colour of face, limbs, and tail as in *D. incana*. Skull and canines shorter than in the latter. (*Thomas*).

Distribution.—Rio Abbaregas, Merida, Venezuela; at an elevation of about 5,000 feet above the sea.

GENUS DROMICIOPS.

Dromiciops, Thomas, Ann. Mag. Nat. Hist., ser. 6, vol. xiii., p. 187 (1894).

Differs from all other opossums by the short and furry ears, hairy tail, the thick and doubly inflated auditory bullæ of the skull, the shortened canines, and the peculiar form and positions of the incisor teeth.

1. CHILOE ISLAND OPOSSUM. DROMICIOPS AUSTRALIS.

Didelphys australis, Philippi, Arch. Naturgesch., 1894, vol. i., p. 33.
Dromiciops gliroides, Thomas, Ann. Mag. Nat. Hist., ser. 6, vol. xiii., p. 187 (1894).

Characters.—Size and superficial appearance very similar to those of the Tasmanian Dormouse-Phalanger (*Dromicia nana*). General colour of upper-parts fawn-grey, darker on back than on sides; fur coloured as in *Didelphys elegans*, that is, pale grey, with distinct black rings round the eyes. Ears short and rounded, with the anterior basal prominence very slightly developed: covered on the back with thick fur like that of the crown of the head, and thus very different from the almost naked ears of the sub-genus *Marmosa*. Crown and nape rufous-brown or cinnamon, lighter on sides of neck, where the hairs are ringed with white near the tip. A large whitish patch behind the shoulder, a second in front of the same, and a third behind the hip. Under-parts dirty yellowish white, with the basal grey of the hairs showing through. Outer side of limbs and back of fore-feet brown; inner side and back of hind-feet dull white. Soles of latter with five prominent, transversely striated pads : a long one at base of hallux, three at the bases of the other toes, and a rather small postero-external pad. Terminal digital pad large, longer than claws, longitudinally striated. Tail very thick at base, and tapering rapidly and evenly to the tip; its basal third thickly clothed with shining fawn-coloured fur like that of the body; its terminal two-thirds almost equally well covered, but the hairs straighter and nearly uniform dark brown. Below, the hairs brownish white throughout. Naked portion of tail confined to narrow strip about half-an-inch in length on the under side of the tip.

Habitat.—S. Chili and Chiloe Island.

FOSSIL FORMS.

Since the text was written a considerable number of new fossil Marsupials have been described, while certain extinct Patagonian forms have been definitely assigned to the group. Space admits of only a brief reference to some of the most important.

In the *Macropodidæ*, De Vis[*] has described several new species of *Palorchestes*, *Sthenurus*, and *Macropus* (*Halmaturus*) from the Pleistocene of Queensland. A new *Macropus* has

[*] Proc. Linn. Soc., N. S. Wales, vol. x., pp. 57—133 (1895).

likewise been recorded by Broom * from the Pleistocene of New South Wales. Most interesting is a small form described by the latter writer † under the name of *Burramys parvus*, which appears to connect *Hypsiprymnodon* very closely with the Phalangers. It has grooved fourth premolars, and two small teeth between the third premolar and the first lower incisor, while there is no pit on the outer side of the lower jaw. He also records remains of the existing *Petaurus breviceps* and *Dromicia nana* from the New South Wales caverns. *Palæope taurus*, from the same deposits, appears to be allied both to *Petaurus* and *Gymnobelideus*. A new species of *Pseudochirus* is likewise recorded by the same writer from these deposits. In the *Diprotodontidæ*, De Vis ‡ considers that *Nototherium* is not entitled to form a family by itself. In the *Phascolomyidæ* the identity of *Sceparnodon* with *Phascolonus* is confirmed by Stirling.§

The reference of the *Epanorthidæ*, *Abderitidæ*, and *Garzoniidæ* of the Tertiaries of Patagonia to the Diprotodontia has been already referred to under the heading of *Cœnolestes*. Whereas in the first of these families the last premolar is of normal form, in the second it is greatly enlarged and grooved.

In the *Peramelidæ*, Broom ‖ records remains of two living species of *Phascologale* from the New South Wales caverns. In the *Dasyuridæ*, De Vis ¶ has described an extinct species of *Thylacinus* from the Queensland Tertiaries, and Ameghino ** has recorded various allied forms from the Santa Cruz beds of Patagonia under the names of *Prothylacinus*, *Napodonictis*, *Anutherium*, *Cladosictis*, *Amphiproviverra*, *Sipalocyon*, and *Ictioborus*.

---

* Proc. Linn. Soc., N. S. Wales, 1896, pp. 48—61.   † *Op. cit.*, p. 563.
‡ *Op. cit.*   § "Nature," vol. 50, pp. 184—188, and 206—209 (1895).
‖ *Op. cit.*   ¶ Bol. Acad. Cordoba, vol. xiii., pp. 380-396 (1894).
** *Op. cit.*

ALPHABETICAL INDEX.

acanthion, Echidna, 242
Acanthoglossus bruijnii, 246
Acrobates pulchellus, 119
  pygmæus, 117, 118
aculeata, Echidna, 241
  Myrmecophaga, 241
aculeatus, Tachyglossus, 241
Æpyprymnus, 66
Æpyprymnus, 71
  rufescens, 71, 72
affinis, Didelphys, 205
  Perameles, 145
agilis, Didelphys, 211
  Grymæomys, 211
  Halmaturus, 32
  Macropus, 32
  Ornithorhynchus, 279
alba, Phalangista, 84
albertisi, Dactylopsila, 103
  Phalangista, 98
  Pseudochirus, 98
albipes, Phascologale, 179
  Antechinus, 179
albiventris, Didelphys, 197
alboguttata, Microdelphys, 221
albopunctatus, Dasyurus, 165
altus, Macropus, 255
  Phascolagus, 255
Amblotherium, 276
  soricinum, 276
americana, Didelphys, 219
americanus, Sorex, 219
Amphilestes, 276
  broderipi, 276
Amphiproviverra, 268

Amphiproviverra, manzaniana, 269
Amphitherium, 277
anak, Macropus, 257
  Protemnodon, 257
anatinus, Ornithorhynchus, 231
  Platypus, 231
angustivittis, Phalangista, 103
annulicauda, Onychogalea, 49
antæus, Protemnodon, 257
Ant-eater,
  Banded, 184
  Marsupial, 182
  Porcupine, 240
  Spiny, 240
Antechinomys laniger, 182
Antechinus albipes, 179
  apicalis, 167
  concinnus, 172
  ferrugineifrons, 180
  flavipes, 172
  froggatti, 178
  fuliginosus, 179
  leucogenys, 180
  leucopus, 180
  maculatus, 174
  minimus, 172
  minutissimus, 174
  moorei, 171
  murinus, 179
  niger, 171
  rolandensis, 172
  stuarti, 172
  swainsoni, 171
antilopinus, Halmaturus, 20
  Macropus, 19

antilopinus, Osphranter, 19
apicalis, Antechinus, 167
　Hypsiprymnus, 65
　Macropus, 25
　Phascologale, 167
archeri, Phalangista, 97
　Pseudochirus, 97
arenaria, Perameles, 137
ariel, Belideus, 105
aruensis, Perameles, 143
atlas, Macropus, 253
　Sthenurus, 253, 254, 257
auratus, Perameles, 144
aurita, Didelphys, 204
australis, Diprotodon, 261
　Echidna, 242
　Petaurus, 106
azael, Palorchestes, 251
Azara's Opossum, 197
azaræ, Didelphys, 197

Banded Ant-eater, 184
Bandicoot, 128
　Broadbent's, 142
　Cockerell's, 143
　Doria's, 142
　Golden, 144
　Gunn's, 139
　Long-nosed, 140
　Long-tailed, 140
　North Australian, 143
　Pig-footed, 146
　Port-Moresby, 145
　Rabbit, 132
　Raffray's, 141
　Short-nosed, 144
　Striped, 137
　True, 135
　White-tailed, 134
banksii, Phalangista, 95
bassii, Phascolomys, 126
beccarii, Dorcopsis, 57
Belideus ariel, 105
　breviceps, 105
　flaviventer, 107
　flaviventris, 107
　gracilis, 105
　notatus, 105
　sciureus, 104

bernsteini, Phalangista, 99
Bettongia, 67
　campestris, 66
　cuniculus, 69
　gaimardi, 69
　gouldi, 70
　grayi, 67
　lesueuri, 67
　ogilbyi, 70
　penicillata, 70
　rufescens, 71
billardieri, Halmaturus, 40
　Kangurus, 40
　Macropus, 40
Bolodon, 283
Boriogale magnus, 24
bougainvillii, Perameles, 137
brachyotis, Halmaturus, 46
　Macropus, 45
　Petrogale, 45
brachyura, Didelphys, 216
　Microdelphys, 215
brachyurus, Halmaturus, 42
　Kangurus, 41
　Macropus, 41
　Peramys, 216
braziliensis, Sorex, 219
brehus, Macropus, 256
　Sthenurus, 256
breviauculeata, Echidna, 242
brevicaudata, Didelphys, 216
brevicaudatus, Halmaturus, 42
　Thylogale, 42
breviceps, Belideus, 105
　Didelphys, 197
　Petaurus, 105
　Phalangista, 85
　Thylacinus, 152
brevirostris, Ornithorhynchus, 231
broadbenti, Perameles, 142
broderipi, Amphilestes, 276
browni, Halmaturus, 37
　Macropus, 37
bruijnii, Acanthoglossus, 246
　Echidna, 246
　Proechidna, 246
　Tachyglossus, 246
bruni, Dorcopsis, 56
　Macropus, 56

ALPHABETICAL INDEX.

brunii, Didelphys, 36
Halmaturus, 36
Macropus, 36
bucklandi, Phascolotherium, 276

Caloprymnus campestris, 66
calura, Phascologale, 176
campestris, Bettongia, 66
Caloprymnus, 66
Hypsiprymnus, 66
cancrivora, Didelphys, 197
canescens, Didelphys, 210
Micoureus, 210
Phalangista, 99
Pseudochirus, 99
canina, Phalangista, 91
caninus, Trichosurus, 91
carcinophaga, Didelphys, 197
carnifex, Thylacoleo, 259
castanotis, Chœropus, 146
caudata, Dromicia, 115
caudivolvula, Didelphys, 94
caudivolvulus, Pseudochirus, 95
celebensis, Cuscus, 86
Phalanger, 86
Phalangista, 86
Chætocercus cristicauda, 167
pilicauda, 170
chalmersi, Dorcopsis, 57
Chironectes, 221
minimus, 221
palmata, 221
variegatus, 221
yapock, 222
Chœropus, 146
castanotis, 146
ecaudatus, 146
occidentalis, 146
cinerea, Didelphys, 207
cinereus, Lipurus, 79
Micoureus, 207
Phascolarctus, 79
cockerelli, Perameles, 143
concinna, Dromicia, 112
Petrogale, 47
Phalangista, 112
concinnus, Antechinus, 172
Halmaturus, 47
Heteropus, 47

concinnus, Macropus, 47
conspicillatus, Halmaturus, 52
Lagorchestes, 52
conspicillatus, Macropus, 52
convolutor, Phalangista, 94
cooki, Phalangista, 96
Pseudochirus, 96
cooperi, Macropus, 255
Osphranter, 255
coxeni, Halmaturus, 34
Macropus, 34
crassicaudata, Didelphys, 202
Phascologale, 178
Sminthopsis, 178
crassicaudatus, Metachirus, 202
Podabrus, 178
crassipes, Halmaturus, 33
Macropus, 33
crispus, Ornithorhynchus, 231
cristicauda, Chætocercus, 167
cristicaudata, Phascologale, 167
cuniculus, Bettongia, 69
Hypsiprymnus, 69
Cuscus celebensis, 86
maculatus, 83
orientalis, 84
ornatus, 86
ursinus, 81
Cuscus, 80
Black, 81
Celebean, 86
Grey, 84
Spotted, 82
Wallace's, 86
cynocephala, Didelphys, 151
cynocephalus, Dasyurus, 151
Thylacinus, 151

Dactylopsila, 102
albertisi, 103
angustivittis, 103
palpator, 103
trivirgata, 103
Dasyure, 150
Black-tailed, 164
Common, 161
North Australian, 165
Papuan, 165
Slender, 160

X 2

Dasyure, Spotted-tailed, 153
  Ursine, 153
Dasyurus, 157
  albopunctatus, 165
  cynocephalus, 151
  fuscus, 166
  geoffroyi, 164
  gracilis, 160
  guttatus, 162
  hallucatus, 165
  laniarius, 267
  macrourus, 158
  maculatus, 158
  maugei, 162
  penicillatus, 175
  tafa, 175
  ursinus, 154
  viverrinus, 161, 162
Dendrolagus, 58
  dorianus, 61
  inustus, 60
  lumholtzi, 58
  ursinus, 60
  derbiana, Didelphys, 206, 207
  derbianus, Halmaturus, 39
Devil, Tasmanian, 154
dicrura, Didelphys, 205
Didelphys, 196, 271
  affinis, 205
  agilis, 211
  albiventris, 197
  alboguttata, 221
  americana, 219
  aurita, 197
  azaræ, 197
  brachyura, 216
  brevicaudata, 215
  breviceps, 197
  brunii, 36
  cancrivora, 197
  canescens, 210
  carcinophaga, 197
  caudivolvula, 94
  cinerea, 207
  cynocephala, 151
  crassicaudata, 202
  derbiana, 206
  dicrura, 205
  dimidiata, 215

Didelphys, domestica, 217
  dorsigera, 209
  elegans, 214
  glirina, 216
  grisea, 212
  gypsorum, 273
  henseli, 218
  hunteri, 216
  iheringi, 220
  impavida, 209
  incana, 212
  karkinophaga, 197
  lanigera, 206
  lepida, 211
  leucotis, 197
  macroura, 202
  marsupialis, 196
  minima, 221
  murina, 209
  musculus, 209
  myosurus, 201
  nana, 211
  noctivaga, 207
  novæ-hollandiæ, 94
  nudicaudata, 201
  ochropus, 206
  opossum, 200
  orientalis, 84
  ornata, 206
  pæcilotis, 197
  penicillata, 175
  peregrinus, 94
  petaurus, 106
  philander, 205
  pusilla, 211
  pygmæa, 117
  quica, 200
  scalops, 217
  sciurea, 104
  sorex, 218
  tricolor, 216
  tridactyla, 65
  trilineata, 219
  tristriata, 219
  turneri, 202
  unistriata, 220
  ursina, 125, 154
  velutina, 213
  virginiana, 197

ALPHABETICAL INDEX.

Didelphys, viverrinus, 161
  volans, 101
  vulpecula, 88
  vulpina, 88
  waterhousei, 207
dimidiata, Didelphys, 215
Diprotodon australis, 261
Distœchurus, 116
  pennatus, 116
domestica, Didelphys, 217
Dorca Kangaroo, 55
  Brown, 56
  Grey, 56
  Macleay's, 57
Dorcopsis, 55
  beccarrii, 57
  bruni, 56
  chalmersi, 57
  luctuosa, 56
  macleayi, 57
  mülleri, 56
doreyana, Perameles, 142
doriæ, Phascologale, 170
dorianus, Dendrolagus, 61
Dormouse-Phalanger, 111
  Common, 113
  Lesser, 115
  Long-tailed, 115
  Western, 112
dorsalis, Halmaturus, 29
  Macropus, 29, 30
  Phascologale. 1, 0
dorsigera, Didelphys, 209
dorsigerus, Micoureus, 209
Dromatherium, 277
Dromicia, 111
  caudata, 115
  concinna, 112
  frontalis, 118
  lepida, 115
  nana, 113
  neilli, 112
  unicolor, 113
Dryolestes, 277
Duck-bill, 230, 231
Duck-Mole, 232, 231

ecaudatus, Chœropus, 146

Echidna, 240, 279
  Black-spined, 247
  Bruijn's, 246
  Common, 241
  Five-clawed, 241
  Hairy, 242
  Port-Moresby, 242
  Three-clawed, 245
Echidna, acanthion, 242
  aculeata, 241
  australis, 242
  breviaculeata, 242
  bruijnii, 246
  hystrix, 241
  lawesi, 242
  nigro-aculeata, 247
  oweni, 279
  ramsayi, 279
  setosa, 242
Echinopus setosus, 242
Egg-laying Mammals, 225
elegans, Didelphys, 214
  Micoureus, 214
eugenii, Halmaturus, 39
  Kangarus, 39
  Macropus, 39

fasciata, Perameles, 137
fasciatus, Halmaturus, 61
  Kangurus, 61
  Lagorchestes, 61
  Lagostrophus, 61
  Macropus, 61
  Myrmecobius, 184
ferragus, Macropus, 255
  Pachysiagon, 255
ferrugincifrons, Antechinus, 180
flavipes, Antechinus, 172
  Phascologale, 172
flaviventer, Belideus, 107
  Petaurus, 106
flaviventris, Belideus, 107
  Petaurus, 106
Flying Phalanger, 104
  Lesser, 105
  Papuan, 106
  Pigmy, 119
  Squirrel, 104

310   ALPHABETICAL INDEX.

Flying Phalager, Taguan, 101
  Yellow-bellied, 106
forbesi, Pseudochirus, 100
formosus, Hypsiprymnus, 69
fossor, Phascolomys, 126
  Wombatus, 125
frænatus, Halmaturus, 50
frænatus, Macropus, 50
frenata, Onychogale, 50
froggatti, Antechinus, 178
  Podabrus, 178
frontalis, Dromicia, 118
fuliginosus, Antechinus, 179
  Trichosurus, 88
fusca, Phascolomys, 126
fusciventer, Perameles, 143
fuscus, Dasyurus, 166
  Ornithorhynchus, 231
  Phascolarctus, 79

gaimardi, Bettongia, 69
  Hypsiprymnus, 69
  Kangurus, 69
garagassi, Perameles, 143
gazella, Halmaturus, 34
geoffroyi, Dasyurus, 164
gigantea, Verboa, 15
giganteus, Macropus, 15
gigas, Phascolomys, 265
  Phascolonus, 265
gilberti, Hypsiprymnus, 64
  Potorous, 64
gliriformis, Phalangista, 113
glirina, Didelphys, 216
goliah, Macropus, 253
  Procoptodon, 253
gouldi, Bettongia, 70
gracilis, Belideus, 105
  Dasyurus, 160
  Macropus, 37, 39
grayi, Bettongia, 67
  Hypsiprymnus, 67
greyi, Halmaturus, 29
  Macropus, 29
grisea, Didelphys, 212
grisescens, Phalangista, 99
griseus, Micoureus, 212
gunni, Perameles, 139
guttatus, Dasyurus, 162

Gymnobelideus, 110
Gymnobelideus, leadbeateri, 111
gypsorum, Didelphys, 273

hallucatus, Dasyurus, 165
Halmaturus agilis, 32
  antilopinus, 20
  billardieri, 40
  brachyotis, 46
Halmaturus brachyurus, 42
  brevicaudatus, 42
  browni, 37
  brunii, 36
  concinnus, 47
  conspicillatus, 52
  coxeni, 34
  crassipes, 33
  dama, 39
  derbianus, 39
  dorsalis, 29
  eugenii, 39
  fasciatus, 61
  frænatus, 50
  gazella, 34
  greyi, 29
  hirsutus, 54
  houtmanni, 39
  inornatus, 46
  irma, 32
  jardinii, 33
  lateralis, 44
  leporoides, 53
  luctuosus, 56
  lunatus, 51
  manicatus, 32
  parma, 40
  parryi, 30
  penicillatus, 43
  ruficollis, 27
  stigmaticus, 35
  temporalis, 36
  thetidis, 38
  thetis, 38
  ualabatus, 24
  unguifer, 49
  wilcoxi, 36
Hapalotis, 181
Hare-Wallaby, 52
  Common, 53

Hare-Wallaby, Rufous, 54
  Spectacled, 52
harrisi, Thylacinus, 151
harveyi, Perameles, 67
Hemibelideus lemuroides, 93
henseli, Didelphys, 218
Heteropus concinnus, 47
herbertensis, Phalangista, 93
  Pseudochirus, 93, 94
hirsutus, Halmaturus, 54
  Lagorchestes, 54
  Macropus, 54
houtmanni, Halmaturus, 39
hunteri, Didelphys, 216
Hypsiprymnodon, 73
  moschatus, 73
Hypsiprymnus, 64
  apicalis, 65
  campestris, 66
  cuniculus, 69
  formosus, 69
  gaimardi, 69
  gilberti, 64
  grayi, 67
  lesueuri, 67
  melanotis, 71
  micropus, 64
  murinus, 65
  ogilbyi, 70
  penicillatus, 70
  phillippii, 69
  platyops, 64
  whitei, 69
hystrix, Echidna, 241
  Ornithorhynchus, 241

iheringi, Didelphys, 220
  Peramys, 220
impavida, Didelphys, 209
incana, Didelphys, 212
inerme, Nototherium, 264
inornata, Petrogale, 46
inornatus, Halmaturus, 46
  Macropus, 46
inustus, Dendrolagus, 60
irma, Halmaturus, 32
  Macropus, 32
isabellinus, Macropus. 22

isabellinus, Osphranter, 22

jardinii, Halmaturus, 33
jukesi, Macropus, 3

Kangaroo, 11
  Antilopine, 19
  Dorca, 55
  Great Grey, 15
  Great Red, 22
  Isabelline, 22
  Musk, 73
  Owen's, 24
  Rat, 63
  Tree, 58
  Wallaroo, 20
Kangurus, billardieri, 40
  brachyurus, 41
  eugenii, 39
  fasciatus, 51
  gaimardi, 69
  laniger, 22
  penicillatus, 43
  ruficollis, 26
  rufogriseus, 27
  rufus, 22
  ualabatus, 24
karkinophaga, Didelphys, 197
Koala, 78, 79
koala, Phascolarctos, 79

lævis, Ornithorhynchus, 231
Lagorchestes, 52
  conspicillatus, 52
  fasciatus, 61
  hirsutus, 54
  leichardti, 53
  leporoides, 53
Lagostrophus, 61
Lagostrophus fasciatus, 61
lagotis, Macrotis, 132
  Peragale, 132, 133
  Perameles, 132, 133
laniarius, Dasyurus, 267
  Sarcophilus, 267
laniger, Antechinomys, 183
  Kangurus, 22

lanigera, Didelphys, 206
  Phascologale, 182
lanuginosa, Phalangista, 95
lasiorhinus, Phascolomys, 128
lateralis, Halmaturus, 44
  Heteropus, 44
  Petrogale, 44
latifrons, Phascolomys, 128
lawesi, Echidna, 242
  Tachyglossus, 242
lawsoni, Perameles, 140
leadbeateri, Gymnobelideus, 111
leichardti, Lagorchestes, 53
lemuroides, Hemibelideus, 93
  Phalangista, 93
  Pseudochirus, 93
lepida, Dromicia, 115
  Didelphys, 211
  Micoureus, 211
leporoides, Halmaturus, 53
  Lagorchestes, 53
  Macropus, 53
lesueuri, Bettongia, 67
  Hypsiprymnus, 67
leucogaster, Phascologale, 173
leucogenys, Antechinus, 180
leucopus, Antechinus, 180
  Phascologale, 180
  Podabrus, 180
  Sminthopsis, 180
leucotis, Didelphys, 197
leucura, Peragale, 134
Lipurus cinereus, 79
longævus, Tritylodon, 282
longicauda, Perameles, 140
longicaudata, Perameles, 141
  Phascologale, 174
Long-snouted Phalanger, 120
luctuosa, Dorcopsis, 56
luctuosus, Halmaturus, 56
lugens, Macropus, 37
lumholtzi, Dendrolagus, 58
lunata, Onychogale, 51
lunatus, Halmaturus, 51
  Macropus, 51
Lutra minima, 221

macleayi, Dorcopsis, 57
Macropus, 14, 254

Macropus, agilis, 32
  altus, 255
  anak, 257
  antilopinus, 19
  apicalis, 25
  atlas, 253
  billardieri, 40
  brachyotis, 46
  brachyurus, 41
  brehus, 256
  browni, 37
  bruni, 56
  brunii, 36
  concinnus, 47
  conspicillatus, 52
  cooperi, 255
  coxeni, 34
  crassipes, 33
  derbianus, 39
  dorsalis, 29
  eugenii, 39
  fasciatus, 61
  ferragus, 255
  frænatus, 50
  giganteus, 15
  goliah, 253
  gracilis, 37, 39
  greyi, 29
  hirsutus, 54
  inornatus, 46
  irma, 32
  isabellinus, 22
  jukesi, 37
  lanigerus, 22
  lateralis, 44
  leporoides, 53
  lugens, 37
  lunatus, 51
  magnus, 24
  manicatus, 32
  minor, 258
  mülleri, 56
  papuanus, 33
  papuensis, 33
  parma, 40
  parryi, 30
  penicillatus, 43
  ræchus, 257
  robustus, 20

ALPHABETICAL INDEX. 313

Macropus, ruficollis, 26
  rufiventer, 40
  rufogriseus, 27
  rufus, 22
  stigmaticus, 35
  thetidis, 38
  tibol, 37
  titan, 254
  ualabatus, 24
  unguifer, 49
  wilcoxi, 36
  xanthopus, 47
Macrotis lagotis, 132
macrotis, Peragalea, 133
  Perameles, 133
macroura, Didelphis, 202
macrourus, Dasyurus, 158
macrura, Perameles, 143
macrurus, Petaurus, 101
  Podabrus, 178
maculata, Phalangista, 82
  Phascologale, 174
  Viverra, 158
maculatus, Antechinus, 174
  Cuscus, 83
  Dasyurus, 158
  Phalanger, 83
magnus, Boriogale, 24
  Macropus, 24
major, Thylacinus, 267
  Triconodon, 274
manicatus, Halmaturus. 32
  Macropus, 32
Manicou, 199
manzaniana, Amphiproviverra, 269
  protoproviverra, 269
Marsupial Ant-eater, 182
  Mole, 187, 189
marsupialis, Didelphys, 196
maugei, Dasyurus, 162
melanotis, Hypsiprymnus, 71
melas, Phascologale, 168
Metachirus crassicaudatus, 202
  myosurus, 201
  opossum, 200
Micoureus canescens, 210
  cinereus, 207
  dorsigerus, 209
  elegans, 214

Micoureus lepida, 211
  murinus, 209
  pusillus, 211
Microdelphys alboguttata, 221
  brachyura, 215
  sorex, 218
  velutina, 213
micropus, Hypsiprymnus, 64
mimas, Protemnodon, 256
minima, Didelphys, 221
  Lutra, 221
  Phascologale, 171, 172
minimus, Antechinus, 172
  Chironectes, 221
  Dasyurus, 171
minor, Macropus, 258
  Sthenurus, 258
minutissima, Phascologale, 174
  Antechinus, 174
mitchelli, Notothcrium, 263
  Phascolomys, 125
  Podabrus, 180
Mole, Duck, 232
Mole, Marsupial, 187, 189
mongan, Pseudochirus, 93
Monotremes, 224
moorei, Antechinus, 171
moresbyensis, Perameles, 145
moschatus, Hypsiprymnodon, 73
mülleri, Dorcopsis, 56
  Macropus, 56
Multituberculata, 280
murina, Didelphys, 65, 209
  Phascologale, 179
  Sminthopsis, 179
murinus, Antechinus, 179
  Hypsiprymnus, 65
  Micoureus, 209
Mus, 181
musculus, Didelphys, 209
Musk-Kangaroo, 73
Myoictis wallacei, 169
myoides, Perameles, 143
myosurus, Didelphys, 201
  Metachirus, 201
  Perameles, 137
Myrmecobius, 182
Myrmecobius fasciatus, 184
Myrmecophaga aculeata, 241

Nail-tailed Wallaby, 48-49
nana, Didelphys, 211
  Dromicia, 113
  Phalangista, 113
nasuta, Perameles, 140
neilli, Dromicia, 112
niger, Antechinus, 171
nigro-aculeata, Proechidna, 247
noctivaga, Didelphys, 207
notatus, Belideus, 105
  Petaurus, 105
Notoryctes, 187
  typhlops, 189
Nototherium, 263
  inerme, 264
  mitchelli, 263
Nototherium victoriæ, 264
novæ-hollandiæ, Didelphis, 94
nudicaudata, Didelphys, 201
nudicaudatus, Pleopus, 74

obesula, Perameles, 144, 145
occidentalis, Chœropus, 146
  Pseudochirus, 96
ochropus, Didelphys, 206
og, Protemnodon, 257
ogilbyi, Bettongia, 70
  Hypsiprymnus, 70
Onychogale, 48
  lunata, 51
  unguifera, 49
Onychogalea, 48
  annulicauda, 49
  frænata, 50
Opossum, 196
  Ashy, 207
  Azara's, 197
  Chilian, 214
  Common, 196
  Grey, 212
  Grey-faced, 217
  Hensel's, 218
  Lesser three-striped, 220
  Metachirus, 200
  Murine, 209
  Philander, 205
  Pigmy, 211

Opossum, Quica, 200
  Rat-tailed, 201
  Red-faced, 217
  Red-flanked, 216
  Shining, 211
  Shrew, 218
  Single-striped, 220
  Tehuantepec, 210
  Thick-tailed, 202
  Three-striped, 219
  Velvety, 213
  Water, 221
  Woolly, 206
  Yellow-flanked, 215
orientalis, Cuscus, 84
  Didelphys, 84
orientalis, Phalanger, 84
ornata, Didelphys, 206
ornatus, Cuscus, 86
  Phalanger, 86
Ornithorhynchus, 231, 279
  agilis, 279
  anatinus, 231
  brevirostris, 231
  crispus, 231
  fuscus, 231
  hystrix, 241
  lævis, 231
  paradoxus, 231
oscillans, Triclis, 258
Osphranter antilopinus, 19
  cooperi, 255
  isabellinus, 22
otuel, Pachysiagon, 253
  Procoptodon, 253
oweni, Echidna, 279

Pachysiagon ferragus, 255
  otuel, 253
pæcilotis, Didelphys, 197
palmata, Chironectes, 221
Palorchestes azael, 251
palpator, Dactylopsila, 103
papuanus, Macropus, 33
  Petaurus, 106
papuensis, Phalangista, 83
  Macropus, 33
paradoxus, Ornithorhynchus, 231

ALPHABETICAL INDEX. 315

parma, Halmaturus, 40
Macropus, 40
parryi, Halmaturus, 30
Macropus, 30
patagonicus, Prothylacinus, 268
penicillata, Bettongia, 70
Didelphis, 175
Petrogale, 43
Phascologale, 175
penicillatus, Dasyurus, 175
Halmaturus, 43
Hypsiprymnus, 70
Kangurus, 43
Macropus, 43
pennata, Distæchurus, 116
Phalangista, 116
pennatus, Distæchurus, 116
Pen-tailed Phalanger, 116
Peragale, 132
lagotis, 132
leucura, 134
Peragalea, 132
Perameles, 135
affinis, 145
arenaria, 137
aruensis, 143
aurata, 144
bougainvillii, 137
broadbenti, 142
cockerelli, 143
doreyana, 142
fasciata, 137
fusciventer, 145
garagassi, 143
gunni, 139
harveyi, 67
lagotis, 132
lawsoni, 140
longicauda, 140
longicaudata, 141
macrura, 143
moresbyensis, 145
myoides, 143
myosurus, 137
nasuta, 140
obesula, 144, 145
raffrayana, 141
rufescens, 143
Peramys brachyurus, 216

Peramys, iheringi, 220
tricolor, 216
tristriata, 219
Peratherium, 272
Peratheutes, 269
pungens, 269
pereginus, Didelphys, 94
Pseudochirus, 94, 95
peronii, Petaurus, 101
Petaurista sciurea, 104
Petauroides, 100
volans, 101
Petaurus Didelphys, 106
Phalangista, 106
Petaurus, 104
australis, 106
breviceps, 105
flaviventer, 106
flaviventris, 106
macrurus, 101
notatus, 105
peronii, 101
pygmæus, 118
sciurens, 104
taguanoides, 101
Petrogale brachyotis, 45
concinna, 47
inornata, 46
lateralis, 44
penicillata, 43
robusta, 20
xanthopus, 47
Petrogalea, 42
Phalanger, 80
Phalanger, 87
Common, 88
D'Albertis', 98
Dormouse, 111
Flying, 104
Forbes's, 100
Herbert River, 93
Hoary, 99
Leadbeater's, 111
Long-snouted, 120
Milne-Edwards', 103
Papuan pigmy, 119
Pen-tailed, 116
Pigmy flying, 117
Ring-tailed 92

Phalanger, Schlegel's 99
  Short-eared, 91
  Sombre, 93
  Squirrel, 110
  Striped, 103
  Yellow, 97
Phalanger celebensis, 86
  maculatus, 82
  orientalis, 84
  ornatus, 86
  ursinus, 81
  volans, 101
Phalangista alba, 84
  albertisi, 98
  archeri, 97
  augustivittis, 103
  banski, 95
  bernsteini, 99
  canescens, 99
  canina, 91
  celebensis, 8
  concinna, 11
  convolutor, 94
  cooki, 96
Phalangista gliriformis, 113
  grisescens, 99
  herbertensis, 93
  lanuginosa, 95
  lemuroides, 93
  maculata, 62
  nana, 113
  papuensis, 83
  pennata, 116
  petaurus, 106
  pygmæa, 117
  rufa, 84
  sciurea, 104
  ursina, 81
  viverrina, 96
  vulpina, 88
Phascolarctus, 78
  cinereus, 79
  fuscus, 79
  koala, 79
Phascolagus altus, 255
Phascologale, 166
  albipes, 179
  apicalis, 167

Phascologale, calura, 176
  crassicaudata, 178
  cristicaudata, 167
  doriæ, 170
  dorsalis, 170
  flavipes, 172
  lanigera, 182
  leucogaster, 173
  leucopus, 180
  longicaudata, 174
  maculata, 174
  melas, 168
  minima, 171, 172
  minutissima, 174
  murina, 179
  penicillata, 175
  pilicauda, 170
  rufogaster, 172
  swainsoni, 171
  thorbeckiana, 168
  virginiæ, 180
Phascolomys, 124, 265
  bassii, 126
  fossor, 126
  fusca, 126
  gigas, 265
  lasiorhinus, 128
Phascolomys latifrons, 128
  mitchelli, 125
  platyrhinus, 125
  ursinus, 125, 126
  vombatus, 126
  wombat, 126
Phascolonus, 265
  gigas, 265
Phascolotherium bucklandi,
Philander virginiana, 197
philander, Didelphys, 205
phillippii, Hypsiprymnus, 69
pilicauda, Chætocercus, 170
  Phascologale, 170
Plagiaulax, 284
platyops, Hypsiprymnus, 64
  Potorous, 64
Platypus anatinus, 231
platyrhinus, Phascolomys, 125
Pleopus nudicaudatus, 74
Podabrus, crassicaudatus, 178

Podabrus, froggatti, 178
  leucopus, 180
  macrurus, 178
  mitchelli, 180
  pæcilotis, Didelphys, 107
Polymastodon, 283
Porcupine Ant-eater, 240
Potorous, 63
  apicalis, 65
  gilberti, 64
  platyops, 64
  rufus, 65
  tridactylus, 65
Pouched Mice, 166
  Brush-tailed, 175
  Chestnut-bellied, 170
  Chestnut-necked, 163
  Common, 179
  Crested-tailed, 167
  Freckled, 167
  Jumping, 182
  Lesser bush-tailed, 176
  Little, 171
  Long-legged, 181
  Long-tailed, 174
  Narrow-footed, 177
  Orange-bellied, 170
  Pigmy, 174
  Red-tailed, 169
  Striped faced, 180
  Swainson's, 171
  Thick tailed, 178
  White-footed, 180
  Yellow-footed, 172
Procoptodon, 252
  goliah, 253
  otuel, 253
  rapha, 253
Proechidna, 245
  bruijnii, 246
  nigro-aculeata, 247
  villosissima, 246
Protemnodon anak, 257
  antæus, 257
  mimas, 256
  og, 257
  ræchus, 257
Prothylacinus patagonicus, 268
Protoproviverra manzaniana, 269

Pseudochirus, 92, 259
  albertisi, 98
  archeri, 97
  canescens, 99
  caudivolvulus, 95
  cooki, 96
  forbesi, 100
  herbertensis, 93, 94
  lemuroides, 93
  mongan, 93
  occidentalis, 96
  peregrinus, 94, 95
  schlegeli, 99
  pulchellus, Acrobates, 119
  pungens, Peratheutes, 269
  pusilla, Didelphys, 211
  pusillus, Micoureus, 211
  pygmæa, Didelphys, 117
    Phalangista, 117
  pygmæus, Acrobates, 117, 118
    Petaurus, 118

Quica, 201
quica, Didelphys, 200

Rabbit-Bandicoot, 132
ræchus, Macropus, 257
  Protemnodon, 257
raffrayana, Perameles, 141
ramsayi, Echidna, 279
  Sceparnodon, 266
Rat-Kangaroo, 63
  Broad-faced, 64
  Brush-tailed, 70
  Common, 65
  Gaimard's, 69
  Gilbert's, 64
  Lesueur's, 67
  Plain, 66
  Prehensile-tailed, 67
  Rufous, 71
  Tasmanian, 69
rapha, Procoptodon, 253
Ring-tailed Phalanger, 94
  Common, 94
  Tasmanian, 96
  Western, 96
robusta, Petrogale, 20
robustus, Macropus, 20

robustus, Petrogale, 20
Rock-Wallaby, 42
  Brush-tailed, 43
  Little, 47
  Plain-coloured, 46
  Short-eared, 45
  West Australian, 44
  Yellow-footed, 47
rolandensis, Antechinus, 172
rostratus, Tarsipes, 120
rufa, Phalangista; 84
rufescens, Æpyprymnus, 71, 72
  Bettongia, 71
  Hypsiprymnus, 72
  Perameles, 143
ruficollis, Kangurus, 26
  Halmaturus, 27
  Macropus, 26
rufiventer, Macropus, 40
rufogaster, Phascologale, 172
rufogriseus, Kangurus, 27
  Macropus, 27
rufus, Kangurus, 22
  Macropus, 22
  Ornithorhynchus, 231
  Potorous, 65

Sarcophilus, 153
Sarcophilus, 267
Sarcophilus laniarius, 267
  ursinus, 154
scalops, Didelphys, 217
  Peramys, 217
Sceparnodon ramsayi, 265
schlegeli, Pseudochirus, 99
sciurea, Didelphys, 104
  Petaurista, 104
  Phalangista, 104
sciureus, Belideus, 104
  Petaurus, 104
setosa, Echidna, 242
setosus, Echinopus, 242
  Tachyglossus, 242
Sminthopsis, 177
  crassicaudata, 178
  leucopus, 180
  murina, 179
  virginiæ, 180
Sorex americanus, 219

Sorex, braziliensis, 219
  Didelphys, 218
  Microdelphys, 218
soricinum, Amblotherium, 276
Spalacotherium tricuspidens, 275
spelæus, Thylacinus, 267
spenseri, Tarsipes, 120
Spiny Ant-eater, 244
Sthenurus, 253
  atlas, 257
  brehus, 256
  minor, 258
stigmaticus, Halmaturus, 35
  Macropus, 35
Striped Phalanger, 102
stuarti, Antechinus, 172
swainsoni, Antechinus, 171
  Phascologale, 171

Tachyglossus aculeatus, 241
  bruijnii, 246
  lawesi, 242
  setosus, 242
tafa, Dasyurus, 175
taguanoides, Petaurista, 101
  Petaurus, 101
Tarsipes, 119
  rostratus, 120
Tasmanian Devil, 154
  ,,   Thylacine, 151
temporalis, Halmaturus, 36
thetidis, Halmaturus, 38
  Macropus, 38
thetis, Halmaturus, 38
thorbeckiana, Phascologale, 168
Thylacine, 151
Thylacinus, 150
Thylacinus breviceps, 152
  cynocephalus, 151
  harrisi, 151
  major, 267
  spelæus, 267
Thylacoleo carnifex, 259
Thylogale, brevicaudatus, 42
tibol, Macropus, 37
titan, Macropus, 254
Tree-Kangaroos, 58
  Black, 60

Tree-Kangaroos, Brown, 60
 Doria's, 61
 Queensland, 58
Trichosurus, 87, 91
 caninus, 91
 fuliginosus, 88
 vulpecula, 88
Triclis oscillans, 258
tricolor, Didelphys, 213
 Peramys, 216
Triconodon, 274
tricuspidens, Spalacotherium, 275
tridactyla, Didelphys, 65
tridactylus, Potorous, 65
trilineata, Didelphys, 219
tristriata, Didelphys, 219
 Peramys, 219
Tritylodon, 282
 longævus, 282
trivirgata, Dactylopsila, 103
turneri, Didelphys, 202
typhlops, Notoryctes, 189

ualabatus, Halmaturus, 24
 Kangurus, 24
 Macropus, 24
unguifer, Halmaturus, 49
 Macropus, 49
unguifera, Onychogale, 49
unicolor, Dromicia, 113
unistriata, Didelphys, 220
Ur-quamata, 189
ursina, Didelphis, 125, 154
 Phalangista, 81
Ursine Dasyure, 153
ursinus, Cuscus, 81
 Dasyurus, 154
 Dendrolagus, 60
 Phalanger, 81
 Phascolomys, 125, 126
 Sarcophilus, 154

variegatus, Chironectes, 221
victoriæ, Nototherium, 264
velutina, Didelphys, 213
 Microdelphys, 213
villosissima, Proechidna, 246

virginiana, Didelphys, 197
virginiana, Philander, 197
virginiæ, Phascologale, 180
 Sminthopsis, 180
Viverra, maculata, 158
viverrina, Phalangista, 96
viverrinus, Dasyurus, 161, 162
 Didelphys, 161
volans, Didelphys, 101
 Petauroides, 101
 Phalanger, 101
vombatus, Phascolomys, 126
vulpecula, Didelphys, 88
 Trichosurus, 88
vulpina, Didelphy, 83
 Phalangista, 88

Wallaby, Agile, 32
 Aru Island, 36
 Banded, 61
 Black-gloved, 32
 Black-striped, 29
 Black-tailed, 24
 Branded, 35
 Bridled, 50
 Cape York, 34
 Crescent, 51
 Dama, 39
 Grey's, 29
 Hare, 52
 Nail-tailed, 48-49
 Pademelon, 38
 Parma, 40
 Parry's, 30
 Red legged, 36
 Red-necked, 26
 Rock, 42
 Rufous bellied, 40
 Short-tailed, 41
 Sombre, 37
wallacii, Myoictis, 169
 Phascologale, 169, 170
Wallaroo, 20
waterhousei, Didelphys, 207
Water-Opossum, 221
whitei, Hypsiprymnus, 69
wilcoxi, Halmaturus, 36
 Macropus, 36

Wombat, 124
  Common, 125
  Hairy-nosed, 128
  Tasmanian, 125
wombat, Phascolomys, 126
Wombatus fossor, 125

xanthopus, Petrogale, 47
  Macropus, 47

Yapock, 221
yapock, Chironectes, 222
Yerboa, gigantea, 15

INDEX TO APPENDIX.

Cœnolestes, 290
  fuliginosus, 292
  obscurus, 292
Cuscus, Short-headed, 288

Dasyuroides, 297
  byrnei, 298
Dendrolagus bennettianus, 287
Didelphys fuscata, 301
  incana, 301
  koseritzi, 299
  marmota, 300
  soricina, 300
  trinitatis, 299
Dromiciops, 302
  australis, 302

Epanorthidæ, 289

Fossil forms, 303

Kangaroo, Bennett's Tree, 287

Mouse, Byrne's Pouched, 298
  Crest-tailed Pouched, 293

Mouse, Larapinta Pouched, 295
  Lesser Brush-tailed Pouched, 294
  Macdonnell's Pouched, 294
  Sand-hill Pouched, 296

Opossum, Chiloe Island, 302
  Grey, 300
  Hoary, 301
  Koseritz's, 299
  Merida, 301
  Shrew-like, 300
  Trinidad, 299

Phalanger breviceps, 288
  Ring-tailed, 288
Phascologale cristicauda, 293
  calura, 294
  macdonellensis, 294
Pseudochirus dahli, 288

Selvas, The, 290
  Bogota, 292
  Ecuador, 292
Sminthopsis larapinta, 295
  psammophilus, 296

www.ingramcontent.com/pod-product-compliance
Lightning Source LLC
Chambersburg PA
CBHW050850300426
44111CB00010B/1198